Autonomous Technologies: Applications That Matter

Other SAE books of interest

Automotive 2030—North America
By Bruce Morey
(Product Code: T-127)

Autonomous Vehicles for Safer Driving
By Ronald K. Jurgen
(Product Code: PT-158)

V2V/V2I Communications for Improved Road Safety and Efficiency
By Ronald K. Jurgen
(Product Code: PT-154)

Autonomous Technologies: Applications That Matter

Edited by William Messner

INTERNATIONAL™

Warrendale, Pennsylvania
USA

SAE INTERNATIONAL

400 Commonwealth Drive
Warrendale, PA 15096
E-mail: CustomerService@sae.org
Phone:
+1.877.606.7323 (inside USA and Canada)
+1.724.776.4970 (outside USA)
Fax: +1.724.776.0790

SAE Order Number JPF-AUV-004
DOI 10.4271/JPF-AUV-004

Library of Congress Cataloging-in-Publication Data
Autonomous technologies : applications that matter / edited by Dr. William Messner.
pages cm
Includes bibliographical references.
Summary: "In a unique effort to offer an introduction to the world of autonomous technologies, SAE International and AUVSI are pleased to present the reader with Autonomous Technologies: Applications That Matter. Focusing on nonmilitary use of autonomous mobile devices, the two organizations have teamed up and invited authors from multiple areas of expertise to discuss how the concept of autonomy is becoming a more natural fit with the way we live"--Foreword.
ISBN 978-0-7680-7718-6
1. Autonomous robots. 2. Automobiles--Automatic control. 3. Agricultural machinery--Automatic control. 4. Remote submersibles. I. Messner, William C., editor.
TJ211.495.A986 2014
629.8'932--dc23

2014018432

ISBN-Print 978-0-7680-7718-6
ISBN-PDF 978-0-7680-8135-0
ISBN-epub 978-0-7680-8137-4
ISBN-prc 978-0-7680-8136-7

Autonomous Technologies: Applications That Matter is a co-publication with SAE International and the Association for Unmanned Vehicle Systems International, 2700 South Quincy St, Suite 400, Arlington, VA 22206.

Table of Contents

Foreword

In a unique effort to offer an introduction to the world of autonomous technologies, SAE International and AUVSI are pleased to present the reader with *Autonomous Technologies: Applications That Matter.*

Focusing on nonmilitary use of autonomous mobile devices, the two organizations have teamed and invited authors from multiple areas of expertise to discuss how the concept of "autonomy" is becoming a more natural fit with the way we live.

From the use of domestic and commercial robots to do tasks such as cleaning and grass mowing, to automated farming, maritime applications, and environmental monitoring (which affects all of us), *Autonomous Technologies: Applications That Matter* looks at the new opportunities and challenges we are already dealing with: Can we trust driverless cars? And who is responsible for keeping the boundaries on privacy issues clear and straight? Will we become more efficient if we embed autonomous platooning in moving cargo and freight?

In addition to gaining efficiency, these technologies also have powerful market incentives. In 2013, a study by Morgan Stanley indicated that self-driving cars could contribute a staggering $5.6 trillion in annual savings around the world. An AUVSI study, also published in 2013, found that unmanned aircraft, employed in the agriculture sector, could create more than 100,000 jobs in the first decade after airspace integration, with more than $82.1 billion in revenue in the United States alone.

Interestingly enough, most of these "autonomy angles" touch upon one common thread: the exponentially growing world of sensors. Ever more precise and dedicated sensors are making data acquisition and decision support much easier and more informed, too.

The scope of this book is practical rather than theoretical, opening the door to future contributions to complement it.

For now, we kindly thank our authors and reviewers for the work and commitment to this project and willingness to share their knowledge.

We hope the readers will find *Autonomous Technologies: Applications That Matter* useful, easy to read, and a first, positive step into the realm of autonomous technologies.

Sincerely,

SAE International
AUVSI

Introduction

William Messner, Tufts University

> My interest is in the future because I am going to spend the rest of my life there.
>
> —Charles Kettering

What a difference 10 years make! When the U.S. government's Defense Advanced Research Project Agency (DARPA) announced its "DARPA Grand Challenge" in 2002, offering a U.S. $1 million prize for an autonomous vehicle to drive over 100 miles across the Mojave Desert in less than 10 hours, the task was widely considered to be impossible. At the first running of the event in March 2004, none of the entrants came close to completing the course. A decade later, not only have autonomous vehicles met the DARPA Grand Challenge and the DARPA Urban Challenge, but there are credible predictions by major automotive manufacturers of the widespread deployment of autonomous vehicles by 2025, and perhaps even earlier. The rapid advances in technology and quick acceptance by consumers have changed the conversation from "if" to "when," with respect to automated civilian driving.

The DARPA Challenges galvanized interest in autonomous cars, making them a realistic possibility in the mind of the public, but autonomous and unmanned vehicles have been increasingly employed in many roles on land, in the water, and in the air over the last 20 years. Military applications have received a great deal of attention, with weaponized unmanned aircraft (drones) being the most prominent. However, unmanned vehicles with varying degrees of autonomy already have many civilian applications. Some of these are quite familiar (such as the Roomba autonomous vacuum cleaner), while others remain largely out of the public eye (such as autonomous farm equipment). Additional applications and more capable vehicles will be rapidly coming to the markets in the years ahead. This book examines a number of economically important areas in which unmanned and autonomous vehicles (also understood here as autonomous technologies) are already in use or soon will be.

The intended audience of this book is the technical manager (and curious consumer) who must make decisions about the development of products by their company or the purchase of equipment and systems, or who anticipate changes in their industry. SAE International and AUVSI, co-publishers of this title, believe that the chapters, briefly described below, provide the information that will assist the reader in identifying profitable opportunities and avoiding costly misconceptions with respect to civilian applications of autonomous vehicle technologies. *Autonomous Technologies: Applications That Matter* brings together chapters on how air, water, and ground vehicles are becoming ever more used and appreciated.

Since many civilian applications are on land, it is natural that the majority of the chapters pertain to ground vehicles. There are separate chapters on passenger cars, commercial freight hauling, agriculture, and mowing and landscaping that we believe will give the reader a good idea of the state of the art and of developments likely to occur in these areas.

Several factors have contributed to this new era of autonomous vehicles and their supporting technologies. First and foremost are the improvements in microprocessors. Computing power continues to double about every 18 months (often known as Moore's Law), which corresponds to computing speed increasing by a factor of about 100 in ten years. Decision, guidance, and navigation tasks that required 10 seconds of computation in 2003 were completed in one-tenth of a second in 2014—instantaneously by the standards of human perception.

Another factor is that solid-state memory, which is physically far more robust than spinning-disk memory, can now store gigabytes of operating systems, programs, maps, and other information at a cost of just a few dollars. Durable sensors suitable for autonomous vehicles are now available or have dropped in price.

A final factor not to be underestimated is the virtuous cycle of improved capabilities leading to increased interest and more resources being devoted to building and improving these systems.

Without a variety of sensors for detecting obstacles, for locating vehicles in space, and for measuring various internal states, unmanned systems would be impossible. The development of lower-cost, robust sensors by auto industry suppliers is the third major type of hardware improvement (joining computation and memory) responsible for the changed outlook in autonomous ground vehicles over the last ten years. Greg Kogut of the U.S. Navy discusses exteroceptive sensors (sensors measuring the exterior environment of a vehicle, such as radar) and proprioceptive sensors (sensors measuring the internal state of a vehicle, such as accelerometers) in terms of what is available and what is to come. His chapter on sensors is relevant to all ground vehicles, but particularly to road vehicles.

Autonomous driving of passenger cars represents the biggest potential economic impact of autonomous vehicle technologies, with a market of hundreds of millions of vehicles, and the prevention of millions of accidents, hundreds of thousands of injuries, and tens of thousands of deaths every year. Already, many vehicles are equipped with driver-assist devices (parking assist and adaptive cruise control are routinely available). In the chapter "Automated Driving: For Real This Time?," Richard Bishop discusses the state of the art and the prospects for passenger cars of varying levels of autonomy from a technological, economic, and regulatory perspective in the United States, Europe, and Asia.

Hauling of freight and cargo, including passengers, is another application of autonomous vehicles. As with cars, various levels of autonomy are likely to make their appearance in the coming years. Although autonomous haulers will be sharing public roads and highways with passenger cars, the value proposition is very different. For commercial enterprises the primary attraction of autonomy is the more efficient use of limited infrastructure. Autonomous vehicles can lead to higher throughput on highways with better lane keeping or by making docking operations more efficient—the backing up of trailers to loading docks or bringing buses to loading stations. In the chapter "Autonomous Vehicle Technologies and the Applications: Cargo and Freight," Dan Williams expands on the value creation, technologies, and vehicle applications for autonomy for commercial vehicles, including semi-tractors pulling trailers and touring coaches that operate primarily in the lane-keeping duty cycle, and urban buses, delivery vans, and refuse haulers, which experience many starts, stops, and turns.

While civilian transportation on roads is where autonomous vehicles have the biggest economic *potential*, agriculture is where autonomous vehicles are having the largest impact *right now*. In fact, major manufacturers of farm equipment have been selling products with driver-assist capability since global navigation satellite systems (for example, GPS) became available for civilian use in the 1990s. Farms are excellent places for the use of autonomous vehicles because they are benign environments compared with roads and highways. There are few people present; farms are highly structured; and farmlands often have unobstructed views of navigation satellites. Farm equipment exploiting such navigation systems can greatly reduce fertilizer, pesticide, and herbicide use while increasing yield by applying these chemicals only where and when they are needed. Nine authors, led by Dr. Marcel Bergerman, from industry and academia contributed to the chapter "Agricultural Applications," in which they convincingly make the case for automation from economic and food security perspectives, as well as covering existing and future technologies.

From a more domestic perspective, the millions of Roomba robot vacuum cleaners dwarf the number of all other robots combined. That product and other autonomous cleaning robots are indeed unmanned vehicles, albeit quite limited in what they can do and where the can go. However, robots for household applications are going to be an area of rapid growth in the coming years, and the capabilities of these systems are rapidly improving. Dr. Bryan Yamauchi of iRobot discusses the opportunities and challenges of handling routine but complex tasks such as cleaning in "Beyond Rosie: Consumer Robots in the 21st Century." One particularly intriguing consumer robot application is telepresence, a mobile extension of what we know as video conferencing. In the future, telepresence will extend beyond mobility to certain manipulation and lifting tasks for assisting the disabled or elderly.

An enticing application of autonomy that superficially resembles farm work, but actually has more in common with consumer robots for household cleaning, is commercial and residential landscaping. There already exist products akin to vacuuming robots for mowing applications for the first adopter/novelty consumer. The real question is not "can it be done?" Rather, it is "what is the business case for doing it?" In the chapter "Landscaping Applications," Alex Foessel-Bunting and Justin Kraft of the John Deere Company examine the technologies, costs, and challenges for producing viable products for commercial and residential landscaping operations. They also examine the value proposition for and against adoption by commercial and residential customers. Mowing is clearly a sweet spot in this space, primarily for releasing the worker or homeowner to attend to more complex tasks, but the authors consider other applications as well.

While land-based unmanned vehicles are primarily intended to replace or augment the operations of existing manned vehicles, unmanned maritime systems are in many cases taking on tasks for which there are no manned systems. More than two-thirds of the Earth's crust lies under oceans and other bodies of water. These regions deeper than 30 meters (100 feet) are largely inaccessible to humans without protective pressure hulls. To get to them with the purpose of developing economic activities and research, remotely operated vehicles (ROVs) and autonomous systems are more and more frequently being employed. A high-profile example of the application of an ROV was the capping of the mile-deep New Horizons wellhead in the BP oil spill disaster. Justin Manley and Thomas Altshuler of Teledyne Maritime Systems examine technologies, opportunities, and challenges of surface and subsurface applications in their chapter, "Unmanned Maritime Vehicles: New Options for Ocean Applications."

Another growing application of autonomous technology in vehicles refers to unmanned aerial vehicles (UAVs), which have existed since the Second World War and come to prominence in the last 15 years for their military applications. However, the many advances leading to greater capabilities for surveillance and reconnaissance by military aircraft also have opened up new possibilities for civilian use. Everett Hinkley of the U.S. Forest Service and Vincent Ambrosia and Steven Wegener of NASA Ames Research Center review the kinds of aircraft and measurements for which various unmanned aircraft are suitable in "Unmanned Aircraft Systems in Environmental Monitoring." They also examine the current and future regulatory climate for unmanned aircraft in civilian air space. Detection of forest fires and response to them is a primary application, but monitoring of tropical storms, air pollution, crops, forest health, and water resources are also viable uses.

The improvements in sensing, computation, and data transmission have made civilian UAVs useful not only for monitoring the environment but also for monitoring members of the public and their activities. Information that is considered appropriate in conflict zones may be a severe invasion of citizens' privacy at home. Brett Davis of AUVSI considers the issues of privacy raised by unmanned aircraft systems (UASs) and the existing laws and court rulings relating to them in "None of Your Business: Unmanned Systems and Privacy." He also reviews prospects for legislation, self-regulation, and public acceptance of UASs.

So, the future is already here when it comes to vehicle autonomy. We will see increasing examples of their applications in the civilian world, helping us be more productive and allowing us to focus on brand-new challenges.

I would like to particularly thank Ms. Monica Nogueira of SAE International for her contributions. In many ways she deserves the title "editor" more than I do. Not only was this book her idea, but also she was instrumental in recruiting several of the authors, and she kept the project going during some difficult times for several authors and myself. She is truly the unsung hero of this volume.

Chapter 1
Sensors

Greg Kogut, U.S. Navy—Space and Naval Warfare Systems Center Pacific

> The senses deceive from time to time, and it is prudent never to trust wholly those who have deceived us even once.
>
> —Rene Descartes

1.1 Introduction

Human-operated vehicles have been designed to facilitate the use of human senses. Autonomous vehicles must replace this dependence with a purely machine-based sensing system, a collection of devices that measure properties of the vehicle and surrounding environment and convert those properties to a form accessible for computation. Machine sensors have some clear advantages, including the ability to measure more quickly, accurately, and reliably. However, matching the performance of human perception in all possible conditions using cost-effective sensors is still an open challenge to the autonomous vehicle research community. Recent progress in research indicates that this goal can be met and surpassed, and that machine sensing can eventually surpass human performance in most areas, including, most importantly, safety of operation.

Autonomous vehicles depend upon sensor data for almost every aspect of driving, including performance, safety, comfort, and reliability; while modern systems already include dozens of sensors, the demands of autonomy dictate even greater use of sensing, and the need for richer, higher-quality, and more reliable data.

Vehicle sensors can be classified into two major categories: proprioceptive and exteroceptive. Proprioceptive sensors measure the internal state of the vehicle, such as inertial sensors to measure acceleration and attitude, and pressure sensors to monitor intake manifold pressure. Exteroceptive sensors acquire information about the environment surrounding the vehicle. Examples of exteroceptive sensors include radar, vision, and lidar (light detection and ranging) (see Figure 1.1). Most proprioceptive sensors are generally technically mature, low cost, and already pervasive in modern vehicles, whereas exteroceptive sensors are rapidly evolving in capability and decreasing in cost. This evolution is vital to the successful development of autonomous vehicles.

Figure 1.1 Autonomy research vehicle showing a variety of exteroceptive sensors, including lidar, radar, and satellite navigation.

1.2 The History of Automotive Sensors

One of the first automotive sensors was the odometer, used to measure distance travelled and instantaneous velocity. The odometer was carried over from horse-carriage odometers pioneered by Benjamin Franklin, among others [1-1]. The use of automotive sensing accelerated through the latter decades of the 20th century with the advancement of electronic design. Examples include the development of the oxygen sensor in 1979 by the Robert Bosch Company [1-2] and the use of accelerometers in the evolution of the modern air bag from the 1950s to the present [1-3]. The oxygen sensor led to modern engine control, playing a significant role in the more than 50% increase in engine efficiency and reduction in vehicle emissions since the 1970s [1-4]. Airbags are now considered standard equipment on all classes of consumer vehicles and have played a major role in the significant increase in vehicle safety over the same period [1-5].

Automotive use of exteroceptive sensors began with the development of adaptive cruise control and parking assistance. Adaptive cruise control uses range sensors to measure the distance to other vehicles on the road, allowing the cruise control to dynamically optimize vehicle speed to maintain a safe distance from the other vehicles. Adaptive cruise control was first marketed by Mitsubishi in 1995 and has since been expanded to many vehicle models by most major auto manufacturers. Ultrasonic sensors have been used since the late 1990s to provide audible proximity warnings to drivers while parking or navigating tight spaces. More recently, parking assistance has expanded to include semiautonomous parallel parking, where the sensor data is used as input to a control system that actuates the steering wheel, leaving the gas and brake pedal for manual operation.

Automotive night vision, or the use of infrared sensing to augment human vision, was first marketed in 2000 and has since expanded to premium models among several manufacturers [1-6].

1.3 Proprioceptive Sensors

Proprioceptive sensors are "inward looking" sensors that measure attributes internal to the vehicle and do not depend upon external signals or properties of the environment. This class of sensor is generally mature, low-cost, and reliable and is pervasive among modern vehicles.

1.3.1 Speed and Position Sensors

Measuring the speed and position of moving parts on a vehicle is a fundamental sensing requirement for both manually and autonomously operated vehicles. Overall vehicle speed is measured either with a single vehicle speed sensor (VSS) mounted near the transmission or transaxle or, increasingly, with anti-lock brake (ABS) speed sensors mounted at each wheel. Other speed and position sensors are also used throughout the vehicle to measure such things as steering-wheel angle, engine RPM, and brake and throttle position. One of the most common types of sensors used to measure angular speed or position is the rotary encoder. Rotary encoders can use mechanical, optical, capacitive, or magnetic properties to detect speed and position. For example, vehicle speed can be calculated by measuring the angular velocity of the four wheel hubs of a vehicle. One of the most common technologies is the Hall-effect sensor, commonly used in ABS brakes. The Hall-effect sensor is a transducer that toggles its output voltage in response to a magnetic field.

1.3.2 Accelerometers

Accelerometers measure linear acceleration and are used in vehicles for safety, navigation, and system diagnosis. For example, accelerometers play a vital role in the automatic deployment of airbags and must operate with high reliability to avoid erroneous activation of the airbag (Figure 1.2). More recently, accelerometers have been used in other safety operations, such as crash detection, allowing for automatic emergency response, as for example the OnStar system produced by General Motors [1-7]. Accelerometers can also be used in on-board diagnostics by measuring abnormal vibration that may be caused by a malfunctioning vehicle process, and they will have increasing use as navigation sensors in autonomous vehicles. Accelerometers alone are not typically as accurate as other standalone navigation sensors, but they are useful when employed in conjunction with other exteroceptive sensors via *sensor fusion*, discussed below.

The accelerometers used in automotive applications are typically microelectromechanical systems (MEMS) devices. These are small, durable, and inexpensive integrated circuits that can easily be incorporated throughout a vehicle as required by application.

Figure 1.2 Airbag control unit with MEMS accelerometers. (courtesy of Continental Aktiengesellschaft)

1.3.3 Yaw-Rate Sensors (Gyroscopic)

Yaw-rate sensors, which measure angular velocity, became commonplace in many modern automobiles through their use in electronic stability control (ESC). In ESC applications, yaw-rate sensors measure angular velocity in the vehicle's yaw axis, the vertical axis through a vehicle's center of gravity. ESC is an application of sensor fusion (discussed below), which combines data quantifying steering-wheel angle, throttle effort, wheel encoders, and other sensors to estimate the stability of a vehicle and enable automatic intervention when appropriate.

Rate sensors are a staple of autonomous vehicles, measuring angular rates in three dimensions, along the roll, pitch, and yaw axes of the vehicle (Figure 1.3). They are used in navigation—tracking the position of the vehicle in the world—as well as providing inputs into the vehicle control system.

Figure 1.3 Fiber optic angular rate gyroscope. (courtesy of KVH Industries)

1.4 Exteroceptive Sensors

Exteroceptive sensors measure properties of the environment surrounding the vehicle and include vision, ranging, and environmental sensors. Exteroceptive sensors are present in many modern manually operated vehicles for such purposes as adaptive cruise control, augmentation of driver vision, and parking assistance, and their use and importance will increase dramatically in the development of full vehicle autonomy as human perception is completely displaced by automated sensing. This increased demand for exteroceptive sensors will likely result in new innovation in the capability and packaging of sensors, as well as a significant decrease in their cost.

1.4.1 Vision Sensors

Vision sensors are imaging sensors that record properties of light. This includes light in the visible spectrum as well as, increasingly, other segments of the electromagnetic spectrum. Vision sensors are closely analogous to human visual sensing, the primary human sense used in manual driving. Vision sensors can be classified into several categories with differing capabilities: monocular, stereo, and infrared.

1.4.1.1 Monocular

Many of today's vehicles use monocular cameras to augment driver vision as back-up or side view cameras (Figure 1.4), with the sensor data typically displayed to the driver through an LED screen. Broad coverage of the area surrounding the vehicle is provided either with wide-angle optics or the "stitching" of multiple camera views into a single image. Autonomous vehicles use monocular camera data for a variety of purposes, including terrain classification, sign detection and reading, lane detection, and the detection of static or dynamic obstacles. In addition, such visual imagery will remain important as visual feedback for vehicle occupants.

Conventional monocular cameras typically detect either grayscale or color imagery. Grayscale imagers detect the intensity of light in a single band of the electromagnetic spectrum, and color cameras use one of several technologies to detect intensity in multiple simultaneous bands. Color imagery contains richer data for use in terrain and other classifiers. However, grayscale imagers are still very useful for their tendency to have better low-light detection, lower noise, higher dynamic range, and higher speed than comparably priced color imagers.

Monocular cameras are generally low in price due to the high volume of production for automotive and other markets. Although cameras are generally small and durable, their optics are susceptible to negative environmental effects from rain, dust, driven sand, and so on, and some supporting technologies may be needed to maintain optimally performing optics under a wide range of conditions.

Figure 1.4 A rearview parking-assistance camera. (courtesy of VW)

1.4.1.2 Stereo

Stereo-vision sensors employ two monocular cameras located at a lateral offset distance from each other. This distance, the stereo baseline, allows the use of binocular vision techniques to calculate the distance to each pixel in the resulting stereo image. The additional depth information is valuable sensor data for many applications, including obstacle detection and terrain classification. Stereo baselines in autonomous vehicles may range from 5 cm to 3 m, with shorter baselines optimized for measuring the distance to objects near the vehicle (e.g., for pedestrian detection) and longer baselines used for longer-distance purposes such as path-planning.

Stereo imagery has strengths and drawbacks relative to the other ranging sensors discussed next. Stereo is typically low cost through use of existing high-volume monocular imagers. Stereo can be used as a passive sensor during the daytime, meaning it can operate in ambient light, requiring no active signal generation as is needed in radar and lidar (see Figure 1.5). Stereo imagery has high spatial resolution relative to lidar and radar, allowing for better interpretation of the size and shape of obstacles. A weakness here is that the accuracy of depth measurements is inversely proportional to the distance to the object being measured. For example, a stereo may be able to measure to an accuracy of a few centimeters at close range, but achieve only an accuracy of meters at its maximum range. While, broadly speaking, range accuracy is more important at close range than long range with respect to autonomous operation, this is still an important drawback relative to radar and lidar. Another weakness of stereo is that it typically only operates well in the daytime or with powerful illumination at night. While infrared cameras can be used to perform stereo in the same manner as visible-light cameras, the high cost of infrared cameras may make their use in stereo cost-prohibitive.

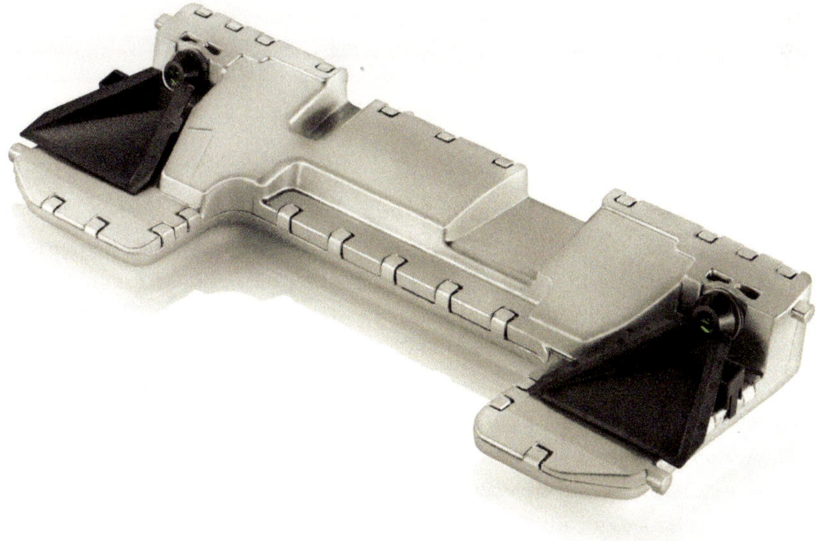

Figure 1.5 An automotive stereo camera. (courtesy of Continental Corporation)

1.4.1.3 *Infrared*

Infrared cameras detect wavelengths of light in the infrared range, or wavelengths longer than detected by human vision. Infrared imagers are currently employed on some manned vehicles to augment human vision with better night vision (see Figure 1.6). In autonomous vehicles, infrared vision is employed in both nighttime and daytime operation and is used for the same applications as visible-light imaging. A strength of infrared imaging is that it tends to be less variant to the time of day or ambient lighting condition, although it is not wholly independent of environmental effects, such as ambient temperature. Drawbacks of infrared vision are its high cost and relatively low resolution compared with visible light cameras. However, the cost of infrared sensors will likely decrease significantly with the increase in volume in automotive applications.

Figure 1.6 Infrared imagery.(courtesy of BMW)

1.4.1.4 *Sonar*

Automotive sonar is active sonar that transmits sound and then detects and receives reflections of the transmitted sound off objects in the environment. The sensor calculates the time interval between transmission and return—called the time of flight—in order to estimate the distance to the objects. Automotive sonar typically uses ultrasonic sound and has ranges of a few centimeters to a few meters. In today's vehicles, sonar is used as a proximity sensor in applications such as parking assist and low-speed obstacle warning (Figure 1.7). Autonomous vehicles will likely use sonar in similar short-range sensing applications and to fill in blind spots left by long-range sensors. Sonar sensors are typically small, low cost, and reliable and may be used throughout the exterior of a vehicle. They are also important as low-speed safety sensors, typically being the last sensor able to detect the range to an impending collision. They also operate in smoke, unlike lidar and vision sensors.

Figure 1.7 Sonar for parking assist. (courtesy of Mazda North American Operations)

1.4.1.5 Lidar

Lidar, similar in principle to sonar and radar, is a distance-measuring sensor that illuminates the surrounding environment with laser light, and then receives the reflected light to calculate distance. Distance is typically calculated through measuring the time of flight of the transmitted laser light. In addition to measuring distance, lidar can also measure some properties of the reflecting surface.

Lidar is attractive as a ranging sensor because of its long range, accuracy, and ability to operate independently of time of day or lighting condition (Figure 1.8). For example, the IBEO *LUX* lidar has a detection range of approximately 0.3 m to 200 m with a distance accuracy of 10 cm, independent of absolute distance [1-8].

Figure 1.8 Autonomous vehicle with a Velodyne HDL-64E scanner lidar. (courtesy of Google, Inc.)

Lidar is emerging among manually operated vehicles as a sensor for adaptive cruise control and obstacle avoidance and for warning applications. In autonomous vehicles, lidar has been used extensively as a sensor for world modeling, obstacle detection and tracking, and terrain classification in autonomy research. Lidar was used by the majority of the vehicles in the DARPA Grand Challenge and Urban Challenge, as well as in the Google self-driving car.

Lidar has several weaknesses relative to other ranging sensors, including sensitivity to some environmental conditions, particularly rain and snow. It tends to have poor spatial resolution and high cost relative to stereo vision sensors. Lidar can also be bulky and challenging to seamlessly integrate with the body of a vehicle. These drawbacks may be mitigated in the future by advances in the development of low-cost flash lidar. Most existing lidar sensors mechanically scan the environment with one or more lasers, but a flash lidar broadly illuminates the entire scene with a pulse of laser light, allowing for higher spatial resolution and potentially higher speed and lower cost. Despite these drawbacks, the overall quality of lidar data will likely result in its extensive use in autonomous vehicles of the future.

1.4.1.6 Radar

Radar is a ranging sensor that transmits and receives radio waves and is similar in concept to lidar. Radar is already emerging in vehicles through its use in adaptive cruise control systems that automatically regulate vehicle speed to maintain a safe distance to vehicles ahead (Figure 1.9). In some systems, the adaptive cruise control can also provide some steering and brake actuation in impending emergency situations, increasing the overall safety of vehicle operation.

In autonomous vehicles, automotive radar will be used for medium- to long-range sensing, with ranges of 0.5 to 200 m, depending upon model and application. Radar has a high update rate relative to stereo and lidar, allowing for quicker response times in collision-detection applications. Like lidar, radar can be operated in any lighting conditions, and it also operates well in most weather conditions. Radar is relatively inexpensive and easy to package compared with lidar, allowing for easier integration with the vehicle body. Some automotive radar has the ability to use the Doppler effect to directly measure the relative velocity of detected objects and also has built-in object tracking algorithms, in some cases simultaneously tracking more than 50 objects [1-9].

A weakness of current automotive radar is relatively poor spatial resolution and field-of-view. Multiple radar units are often necessary to achieve appropriate coverage around an autonomous vehicle. The constraints on field-of-view and spatial resolution may be to some extent mitigated with the development of automotive beam-steering or synthetic aperture radar (SAR).

Despite these weaknesses, automotive radar is ready for use as a reliable, low-cost, and easy-to-integrate sensor.

Figure 1.9 Typical automotive radar module. (courtesy of Delphi Corporation)

1.4.1.7 Satellite Navigation

Satellite navigation is the use of radio-frequency signals generated by multiple satellites to obtain position information. The Global Positioning System (GPS) is already becoming ubiquitous in modern cars, generally with position accuracy of a few meters in typical conditions. In addition to the U.S.-based GPS system, the Russian GLONASS is also globally available; the Chinese *Beidou* and European Union *Galileo* are also scheduled to be globally available in the coming decade. Future satellite navigation sensors may receive signals from multiple satellite systems to improve performance, as well as signals from fixed ground-based reference stations that can improve accuracy to within centimeters.

Although satellite navigation is very useful because it provides an absolute geo-referenced position, it cannot be exclusively relied upon by autonomous vehicles as a position-sensing sensor. Satellite navigation performance can degrade in the presence of high buildings or vegetation that interferes with the satellite transmission. In addition, external interference by other sources of electromagnetic radiation can degrade the signal. To mitigate the effects of these weaknesses, autonomous vehicles typically fuse satellite navigation data with other sensor data to produce a collective estimation of vehicle location. Because of its low cost and generally reliable data, satellite navigation will become a primary sensor on almost all future autonomous vehicles.

1.5 Sensor Fusion

As each sensor discussed has strengths and weaknesses, it is unlikely that any one sensor type will be able to meet the sensing requirements of a fully autonomous vehicle in all conditions. In many cases, the data from multiple types of sensors may be combined by one or more sensor-fusion techniques. For example, research in vehicle autonomy has already resulted in techniques for fusing vision, lidar, sonar, and stereo data into a common coordinate system for use by higher-level algorithms (Figure 1.10). This process involves using the known relative locations of all the sensors on a vehicle as well as the properties of light, optics, and electromagnetic radiation in order to project the data from multiple sensors into a single model of the world.

Figure 1.10 Sensor fusion. The top image shows lidar data, with the bottom showing the lidar data projected into the coordinate system of a camera on the same vehicle. (courtesy of the University of Parma)

Some automotive suppliers are already packaging multiple sensors into integrated packages or providing methods of seamlessly fusing data from multiple sensors of the same type [1-8, 1-10]. The trend toward increased use of fusion is likely to continue given the stringent safety and performance requirements necessary for a fully autonomous vehicle operating in the real world.

1.6 Sensor Processing

The multiple sensors in autonomous vehicles will produce an enormous quantity of data. For example, a single lidar unit can produce millions of bytes of raw data per second. Transmitting and processing this raw data quickly enough for full autonomous operation will be a continuing challenge in the development of autonomy. One strategy to mitigate this problem is to perform much of the processing of raw data in the sensor package itself. For example, some automotive lidar and radar units perform vehicle or obstacle detection and tracking on processors co-located with the sensor [1-8, 1-10]. Many stereo sensors perform the computationally intensive stereo processing algorithms on dedicated digital-signal processing chips located on the stereo head. This type of pre-processing reduces the amount of data necessary to transmit from the sensor and also reduces the computational requirements on the vehicle processors performing higher level autonomy. This trend toward distributed embedded processing will likely continue.

1.7 Conclusion

Autonomous vehicles are sophisticated mobile sensor platforms, carrying dozens of sensors working in concert to produce safe, effective, autonomous driving. With each type of sensor having strengths and weaknesses, no single sensor will be a "silver bullet" for autonomy. Rather, each sensor will contribute and complement the data from the others, producing a rich picture of the vehicle and the surrounding environment. Improvement in sensor technology will correlate directly with the performance and safety of autonomous vehicles and will continue to be a critical area for research and development.

1.8 References

1-1. Benjamin Franklin's Inventions, Discoveries, and Improvements. Retrieved from http://www.ushistory.org/franklin/info/inventions.htm.

1-2. 30 Years of the Bosch Lambda Sensor. Retrieved from http://www.bosch-presse.de/presseforum/details.htm?txtID=2599&tk_id=108.

1-3. Wu, C.-C., et al. (2001). Passive Restraint Control System for Vehicles. U.S. Patent 6272412, filed November 9, 1998, and issued August 7, 2001.

1-4. EPA. (November 2009). Light-Duty Automotive Technology, Carbon Dioxide Emissions and Fuel Economy Trends: 1975 through 2009. Report EPA420-S-09-001. Washington, D.C.: U.S. Environmental Protection Agency, Office of Transportation and Air Quality.

1-5. Olson, C., et al. (2006). Association of First- and Second-Generation Air Bags with Front Occupant Death in Car Crashes: A Matched Cohort Study. *American Journal of Epidemiology* (15 July 2006) 164 (2): 161-169.

1-6. Cadillac Introduces Night Vision Technology. GM. January 16, 2000. Retrieved from http://archives.media.gm.com/ca/gm/en/news/releases/archived%20releases/8f8b67f39a2b5d948525696d00711289.htm.

1-7. The OnStar Automatic Crash Response System. https://www.onstar.com/web/portal/emergencyexplore.

1-8. Ibeo Automotive Standard Four Layer/Multi-Echo LUX Sensor. Retrieved from http://ibeo-as.com.

1-9. Delphi, Safety Electronics, Delphi Electronically Scanning Radar. Retrieved from http://delphi.com/manufacturers/auto/safety/active/electronically-scanning-radar/.

1-10. Delphi, Safety Electronics, Delphi Integrated Radar and Camera System (RACam). Retrieved from http://delphi.com/manufacturers/auto/safety/active/racam/.

Chapter 2
Automated Driving: For Real This Time?

Richard Bishop, Principal, Bishop Consulting

> Personally, I am convinced that the majority of tomorrow's car owners will not even dream of buying a car without autonomous driving possibilities.
>
> —Stefan Jacoby, President and CEO, Volvo Car Corporation

2.1 Introduction

Personal mobility is being transformed by the advent of automated driving. Vehicle automation combined with connectivity also promises to bring unprecedented levels of traffic efficiency. Carmakers are talking about a crashless society, and transportation engineers are talking about the end of congestion–a tall order, indeed: can this really be?

In this decade, we can expect to see the effects of sensor-based active safety systems bringing the crash rate down as these technologies continue to expand across a wider range of vehicle models. Connectivity for infotainment purposes is already a major market driver, and control assist for safety and convenience purposes is laying the groundwork for semiautomated and automated driving. Direct vehicle-vehicle and vehicle-infrastructure connectivity opens up new capabilities.

What is the ultimate in road travel? The ideal road trip should be safe, smooth, uninterrupted, and expeditious. For the vehicle occupants, the time should be productive, restful, and/or entertaining, and they should be aware of information they care about. For most people, it is important that the trip be "good for me and good for society," that is, environmentally friendly. Lastly, accessing this capability needs to be affordable.

Automated driving promises to get us closer than ever to this ideal road trip. For the last few years, the news media has been abuzz with the idea of self-driving cars, and the public is intrigued. A poll conducted by Accenture found that about half of the 2000 individuals surveyed said they "would be comfortable using a driverless car" [2-1]. The J. D. Power 2012 Emerging Technologies Study surveyed over 17,000 vehicle owners and found that 20 percent "definitely would" or "probably would" purchase autonomous driving in their next vehicle–and this was after learning the nontrivial estimated market price of $3000 [2-2.

The demand is there. The days of getting one's first car as a rite of passage are waning, and for the younger generation personal identity is measured more by one's presence on social networks than by make of their automobile. Reduced levels of car ownership could accelerate trends toward more car-sharing, which can be supported by automation.

This chapter briefly examines the history of automated vehicles, including a commentary on how the situation has changed compared with ambitious programs of the past. Terminology is discussed, as are operating scenarios, high-level technical requirements, and the technology basis for automated vehicle systems. Activity and programs from both industry and government are reviewed, leading to a discussion of key issues going forward and expected timing of product introductions.

2.2 History: The Long Quest for the Self-Driving Car

The concept of automated vehicles was originally introduced by General Motors (GM) at the 1939 World's Fair. In the ensuing decades, prototypes were developed in Europe, Japan, and the United States, generally based on inductive sensing of current flowing in wires buried in the pavement for lateral guidance and radar for longitudinal sensing. These systems were evaluated as to performance and cost, remaining in the realm of long-range research.

Beginning in the late 1980s, the European PROMETHEUS project developed an automated vehicle prototype based only on on-board sensors; this work culminated in an automated round-trip from Bavaria to Denmark at highway speeds, with 95% of the driving in autonomous mode (robot car video: http://www.youtube.com/watch?v=AlLZhcnpXYI). During the late 1990s, the European CHAUFFEUR project implemented platooning of heavy trucks and examined related business and societal issues [2-3, 2-4].

In 1992, at the direction of Congress, the U.S. Department of Transportation (USDOT) initiated an ambitious $90M Automated Highway System (AHS) program, partnering with the National Automated Highway System Consortium led by GM. It culminated in the very successful Demo '97, in which thousands of people experienced automated driving on Interstate15 in San Diego (Figure 2.1) [2-3, 2-5, 2-6] (Demo '97 video: http://www.youtube.

com/watch?v=Dtj9J1A-PMk). However, long-range visionary research is especially vulnerable in a tight fiscal environment, and the program was terminated in 1998 due to Congressional budget pressures.

Figure 2.1 NAHSC Cars in platoon formation at Demo '97. (courtesy of UC-Berkeley)

Japanese researchers were continuously active from the 1960s, and in the 1990s the Advanced Cruise-Assist Highway System Research Association (AHSRA) was created by the government and vehicle industry. A variety of automated vehicle techniques were prototyped and demonstrated at their Demo 2000, with an emphasis on infrastructure-supported systems [2-3].

As civilian-focused automation work waned at the turn of the century, attention shifted back to the United States, where the Defense Advanced Research Projects Agency (DARPA) took up the charge with the Grand and Urban Challenges [2-3, 2-7, 2-8]. While the early Challenges attracted mainly academics and roboticists, the 2007 Urban Challenge drew in automotive industry participation. Here, driverless automated vehicles operated among each other in a residential street setting. The top three finishers (Carnegie Mellon University (CMU)/GM, Stanford University, and Virginia Tech) successfully handled traffic circles, four-way stops, parking lots, and road obstacles on these streets (DARPA Urban Challenge Video: http://archive.darpa.mil/grandchallenge/gallery.asp). The expertise developed in the DARPA Challenges flowed into development labs throughout the industry, notably Google's self-driving car team, whose media debut in 2010 fascinated the public and catalyzed new market activity. During this time, nondefense work in the United States was centered at the University of California, Berkeley (UC-Berkeley), which developed and tested close-headway, three-truck platoons in several phases, as well as automated transit bus applications.

In Europe, the focus was mainly on crash avoidance systems during the last decade. However, the Cybercars program in 2001 began developing urban vehicles for automated operation in city centers in lanes or districts segregated from normal traffic, work that has proceeded continuously through today [2-9]. Also, during 2005–2009, the German KONVOI project assessed the impacts of automated truck convoys on surrounding traffic [2-10, 2-11]. A manually driven lead vehicle was followed by three automated followers, with intervehicle gaps on the order of 32–49 feet (10–15 m). Over 1864 miles (3000 km) of testing were conducted on public roads. Passenger car automation work in Europe resumed with the HAVEit project in 2008, discussed later in this chapter.

In parallel with this automation work in the research domain, the auto industry has been introducing a steady stream of active safety systems for crash avoidance since the late 1990s, beginning with warning-only systems and evolving to active intervention [2-3, 2-12] (The Thinking Car documentary: http://www.snagfilms.com/films/title/the_thinking_car). A well-equipped vehicle today may have adaptive cruise control (ACC) providing automatic headway-keeping, automated emergency braking in forward collision situations [2-13], lane departure prevention with active steering (lateral) control, blind spot monitoring, night vision with pedestrian detection, sign recognition, and drowsy driver detection. If this highly capable technology were on all vehicles on the road today, most likely the crash rate would drop precipitously. These features are rapidly moving into the mid-range: a $30,000 car can now be purchased with several of these systems.

Twenty years ago, system engineers debated the proper allocation of intelligence between the vehicle and the highway for implementing vehicle automation. The argument quickly evaporated due to a remarkable increase in vehicle intelligence during the 1990s, which has since accelerated. What else is different? Virtually everything. Now it is the private sector, not government, introducing the technology, and at the most fundamental level, only customer satisfaction has to be delivered (while maintaining safety). This contrasts with the many societal and institutional issues that had to be addressed by government programs. Also, in contrast to some early systems, over a decade of acceptable performance of sensor-based active safety systems has given auto manufacturers the confidence to develop first-generation automated driving systems that do not need support from the roadway in the form of communications or separate lanes for segregation from human-operated vehicles–removing potential barriers that could have resulted from a drawn-out infrastructure deployment process.

From a car industry perspective, the advent of vehicle automation appears to be a given. Further, an exciting new opportunity comes in expanding mobility for the disabled and elderly using vehicle automation, which is especially compelling given the aging of the Baby Boom generation. Google's recent activity acted as the trigger to bring automated driving into the automotive product development realm as a new feature to relieve drivers of the tedium and stress that characterizes the vast majority of driving time. A particularly wrong-minded assertion seen in the media has been to compare Google *prototypes* to automotive *product* and conclude that Google is far ahead of the automotive industry. This is of course an invalid apples-to-oranges type comparison; the auto industry has extensive depth and breadth in this domain, and the major companies have been working quietly on this capability continually for many years.

2.3 Terminology

Careful definitions of terminology are essential to discussions regarding automated vehicles: The term "autonomous vehicle" has frequently been used synonymously, but this is not accurate. The term "automated" is defined as mechanical or electronic devices that function to take the place of human operation. The term "autonomous" simply means that a system or entity is not subject to control from outside elements. Therefore, an autonomous automated vehicle uses only on-board sensors and other electronics to control the vehicle and take over driving. An automated vehicle that makes use of vehicle communications would be considered a cooperative automated vehicle.

Various means of defining levels of automation have been put forth in recent years in both Europe and the United States (Table 2.1). As one example, the U.S. National Highway Traffic Safety Administration (NHTSA) has developed a five-level description, as summarized here [2-14]:

- **Level 0 – No Automation**: The human is in complete and sole control. Examples of driver assistance systems are forward collision warning and lane departure warning.
- **Level 1 – Function-specific Automation**: The driver has complete authority and is solely responsible for safe operation of the vehicle but can cede limited primary control to lateral or longitudinal automation. Examples are adaptive cruise control or lane keeping. Alternatively, an automated system can assume limited authority over a primary control, such as in electronic stability control. The key distinction is that lateral and longitudinal driving support systems are not active simultaneously.
- **Level 2 – Combined Function Automation**: Here, at least two automated primary control functions designed to work in unison relieve the driver of control of those functions. Authority is shared, and the driver is still responsible for monitoring the roadway and safe operation and is expected to be available for control at all times and on short notice. An example would be adaptive cruise control in combination with lane keeping on a limited access highway.
- **Level 3 – Limited Self-Driving Automation**: Here, the driver cedes full control of all safety-critical functions to the automated driving system under certain traffic or environmental conditions, relying on the system for situational monitoring and detection of conditions requiring driver control. When this occurs, the driver is expected to be available to resume control with sufficiently comfortable transition time. The vehicle is designed to ensure safe operation during the automated driving mode. An example would be an automated driving system encountering an oncoming construction area that is not clearly marked and signaling to the driver to reengage in the driving task.
- **Level 4 – Full Self-Driving Automation**: The vehicle is designed to perform all safety-critical driving functions and monitor roadway conditions for an entire trip. The driver provides destination or navigation input but is not expected to be available for control at any time during the trip. This includes both occupied and unoccupied vehicles. By design, safe operation rests solely on the automated vehicle system.

Table 2.1 Levels of Vehicle Automation			
Level	Function	Responsibility	Human Role
Level 0	Normal manual driving	Human	Operator
Level 1	One axis automated, either lateral or longitudinal	Human	Operator
Level 2	Both lateral and longitudinal control automated	Human (ready to take over quickly)	Supervisor
Level 3	Both lateral and longitudinal control automated	Human/Machine (human ready to take over within a comfortable transition time)	"Captain"
Level 4	Both lateral and longitudinal control automated	Machine	Passenger

2.4 Industry Motivation

What is "driving" the automotive industry and other players to pursue automated driving? Strong statements made recently by industry leaders are an indication of how "real" product development of automated driving actually is.

Automated driving is part of Mercedes-Benz's vision for the future of the automobile: they see value in providing greater safety, achieving better traffic flow, relieving the driver of cumbersome routine, and enabling safe use of entertainment/communication devices, with the societal benefits of enabling the elderly and disabled to travel with their own vehicles [2-15]. Similarly, Nissan sees automation happening by 2020, driven by the market opportunity of an aging population with a desire to retain their mobility as they get older [2-16]. Volvo Cars has set a vision that "no one is killed or injured in a new Volvo by 2020," of which automated driving is an important component [2-17, 2-18]. Their research has concluded that younger consumers in particular are willing to pay for technology that can help manage distractions created by the urge to be constantly connected while driving.

Ford's "Blueprint for Mobility" outlines their views on how transportation will evolve in the near-term and beyond [2-19]. The blueprint addresses the technologies, business models, and partnerships needed to tackle the challenges of population growth and urbanization, based on forecasts that show that the 1 billion cars on the road today could quadruple by mid-century. This blueprint notes the importance of automation in saving time, conserving resources, and lowering emissions. Audi has defined their automated vehicle concept as "piloted driving," which takes over the driving task for a limited period of time to allow drivers more comfort during the trip [2-20]. BMW Group has announced plans for highly automated driving functions to be ready for implementation in the 2020 timeframe [2-21]. They are motivated by road safety, citing a "Vision Zero (Accident-Free Mobility)" aim.

Improved convenience and efficiency are additional goals. BMW Group also sees automated driving playing a major part in ensuring sustainable personal mobility in the future.

What about Google? How did a search engine company get into the vehicle technology business? The company has not revealed any specific business models; however, one of the company founders notes that their mission "to create technology that improves the world" includes vehicles [2-22]. Google has also noted that "Automation could save many lives, reduce the nation's energy consumption, free up substantial time, triple the capacity of the highway system, and enable entirely new models of car sharing" [2-23].

The discussion thus far has focused on passenger cars. How do truckers, the other major users of the highway system, view automated driving? While research has been done, engagement by trucking fleets is in its early stages. Yet, they are motivated to reduce fuel consumption and address a perennial driver shortage. Both truck OEMs and fleets are now developing projects for feasibility testing, as described next.

2.5 Technology Basis

The sensing technology to support most automated driving approaches is straightforward, consisting of combinations of camera, radar, and laser scanners, as well as ultrasonics in some applications. Automotive cameras, radars, and ultrasonics are now produced in quantities of millions of units per year. Automotive laser scanners are entering the vehicle marketplace very soon. The cost of these systems is not a barrier to bringing automated driving to the market, particularly since many of the supporting sensors are already engineered into vehicles for crash avoidance purposes. Speaking very generally, and referring to high-volume purchasing, ultrasonic sensors are now in the range of $5 per unit, cameras are well below $100, radars are in the $100 range, and laser scanners are in the low hundreds of dollars.

2.5.1 Camera Sensors

Typical automotive camera modules supplied today use a single imager and image processing techniques for target classification and tracking to support a variety of driver assist functions. The imager itself is a complementary metal oxide semi-conductor high dynamic range (CMOS HDR) camera with high-performance optics (the high dynamic range is needed for sudden lighting changes, such as emerging from a tunnel into bright sunlight). The horizontal field of view is in the range of 45 degrees, with an approximately 30 degree vertical field of view [2-24].

Night vision is another type of imaging used for driver assist (Figure 2.2). Some of the latest systems use data from thermal imagery infrared cameras to detect animals on the road that are beyond the vehicle headlights' field of vision, building on previous versions that can detect pedestrians [2-25].

Figure 2.2 Night vision with pedestrian detection. (courtesy of Mercedes Benz)

One of the newest developments in automotive sensing products is stereo cameras (Figure 2.3). Going beyond monocular camera systems, stereo systems can determine the position and movement of objects along the horizontal, vertical, and longitudinal axes for every image pixel. Further, analyzing differences between the two images enables recognition of objects even when several of them are close together or in low light or poor contrast conditions. This can be helpful in particular for pedestrian detection when there are crowds of people. As an example, the stereo vision system developed by a major automotive supplier has a 197-foot (60–meter) range and can determine the location of objects to within 8–12 inches (20–30 centimeters) from a distance of 66–98 feet (20–30 meters). The cameras are mounted 8 inches (20 centimeters) apart in a housing located behind the windshield [2-26].

Figure 2.3 Example of stereo vision assembly. (courtesy of Bosch)

2.5.2 Radar Sensors

The frequency band of 76–77 GHz is approved for automotive radar applications worldwide. For highway and urban operations, both long- and mid-range radar sensing is needed [2-27, 2-28, 2-29, 2-30]. Long-range sensing (820 feet [250 m]) with a relatively narrow field of view (30 degrees) is used for tracking vehicles and any other in-lane objects at highway speeds. Wide (45 degree) mid-range (525 feet [160 m]) coverage allows detection of vehicles cutting in from adjacent lanes and also identifies vehicles and pedestrians across the width of the equipped vehicle.

To fulfill these sensing requirements, some forward-looking radar systems use multiple beam radars with mechanical scanning, electronic scanning, or several fixed, overlapping beams. Other systems combine a wide field of view at mid-range with long-range coverage to provide two measurement modes simultaneously, using a single radar. Typically, up to 64 targets in the vehicle's region are tracked across multiple lanes. Systems are capable of self-calibration and self-alignment.

2.5.3 Camera-Radar Sensor Fusion

Sensor data fusion is fundamental to forming situational awareness for driver assistance systems. Today's crash avoidance systems typically combine data from radar and vision sensors for greater robustness, as the sensor's strengths and weaknesses complement one another (Figure 2.4) [2-31]. One of the latest crash avoidance systems uses sensor fusion for detecting a bicyclist in the same lane swerving out in front of the car–a dangerous and not unusual occurrence in urban driving. The radar has a wide field of view, and the high-resolution camera allows for robust detection in the complex urban traffic environment [2-32].

Figure 2.4 Camera-radar sensor fusion enables robust detection of objects. (courtesy of Bosch)

2.5.4 Laser Scanners

Similar to radar, laser scanners work according to the principle of time-of-flight technology, enabling identification of objects and other features in the scene to create a dynamic map (with range on each pixel) supporting driver assist applications (Figure 2.5) [2-33]. One approach being manufactured for passenger car application tracks more than 30 objects using a moving integrated mirror that steers the laser beam along the horizontal aperture angle. The sensor's horizontal field of vision is 145°. To account for roadway slopes and vehicle pitch, the laser scanner must operate in multiple vertical planes as well. In our example, each layer covers an aperture angle of 0.8°. This enables a constant assessment of the location of the road surface as a reference point for further processing of the road situation [2-34].

The physics of light-based sensor systems create challenges in adverse weather conditions such as rain, snow, or dusty environments, as airborne particles interfere with sensing. For automotive applications, a multi-echo technology approach is used. Airborne particles reflect some but not all of the energy, so the sensor receives several echoes per laser pulse in order to still get reflections from actual objects. In this way, acceptable object detection can be achieved in most weather conditions. Detection accuracy is on the order of 3.9 inches (10 cm) at a scan frequency of up to 50 Hz (Ibeo Lux Laser Scanner Embedded Object Detection and Tracking: http://www.youtube.com/watch?v=IK29M22teJE).

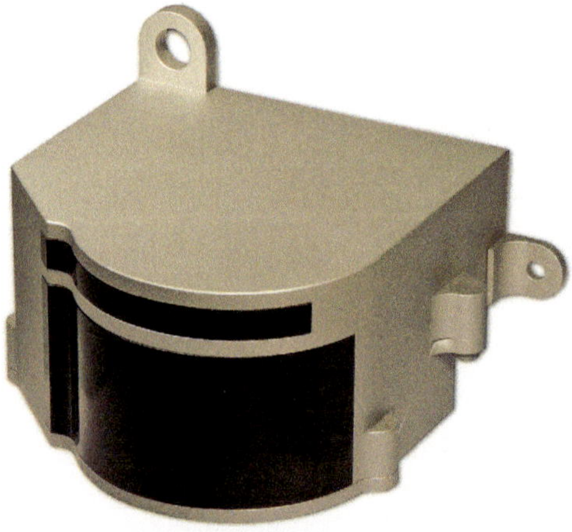

Figure 2.5 Valeo laser scanner for high-volume production. (courtesy of Valeo Automotive Systems)

2.5.5 Low-Latency Vehicle Communications (V2X)

Extensive development of Vehicle-to-Anything (V2X) communications has been ongoing for several years; this is a very broad topic, and only a cursory description is provided here. V2X is important for cooperative safety applications. Early vehicle automation systems do not require V2X, but later generations will likely incorporate it as an additional sensor input. For gaining improvements in traffic flow, V2X and automation are an essential combination.

For safety applications, the premise of V2X is for vehicles to broadcast basic operating data (position, speed, heading, braking status, and so forth) to nearby vehicles (on the order of a 985-foot [300 m] range). Received data from other vehicles can provide greater situational awareness, particularly in cases such as a large vehicle immediately ahead restricting the range of other sensors. The broadcast rate under consideration is from 2 to 10 Hz, depending on the application (general safety advisory versus crash imminent warning). Although the existing 4G cellular network has high performance, a low-latency dedicated channel is required for V2X safety critical communications.

Activity in Europe and the United States relies on a similar spectrum allocation in the 5.9 GHz range. A variety of enabling standards are complete or nearing completion. Activity in Japan is an adaptation of the 5.8 GHz Electronic Tolling standard, and operation in other bands is being considered. In addition to vehicle-to-vehicle (V2V) communications, vehicle-infrastructure (V2I) communications provide connectivity to important infrastructure elements, such as traffic signals, electronic signage, road condition, and detailed road geometry information (Vehicle-to-Vehicle Communication: A New Generation of Driver Assistance and Safety: http://www.its.dot.gov/library/media/v2v_video.htm).

Introduction of V2X-equipped vehicles in the European market is expected beginning in 2015 [2-35]. USDOT is considering a new safety regulation requiring V2V communications [2-36]; if this is the case, V2V would likely be required by late this decade. The Japanese government has already deployed a network of "ITS Spots" along freeways throughout the country and has plans to extend deployment to non-highway roads; in-vehicle equipment is available to receive the data.

2.5.6 Mapping

Google vehicles rely on a database created by a road survey using a vehicle equipped with a high-performance laser scanner; this data is added to highly detailed maps of the roads and terrain. When the automated system is driving the vehicle, it is comparing real-time sensor data to the database for greater robustness in perception [2-37].

2.6 Operational Scenarios and Functional Requirements

2.6.1 Operating Scenarios

Operating scenarios for automated vehicles can be defined as follows: a) limited access highway: a multilane interstate highway, autobahn, and so on, operating at high speeds; obstacles of any sort must be detected; b) rural highway: a single lane each direction operating at high speeds and including at-grade intersections, traffic lights, stop/yield signs, pedestrians, bicycles, and so on; c) urban streets: multilane roads operating at moderate to slow speeds, with frequent intersections, traffic lights, stop/yield signs, pedestrians, bicycles, and so on; d) residential streets: typically a single lane in each direction operating at relatively slow speeds, with traffic lights and a greater prevalence of pedestrians, bicycles, and stop/yield signs; e) parking situations: typical parking lot or multilevel parking garage with pedestrians and other slow-speed vehicle traffic; and f) unimproved roads: gravel or dirt roadways or trails.

2.6.2 Functional Requirements

Functional requirements for automated driving are essentially identical to the functional requirements as we drive our own vehicles, plus further needs directly relating to automation. These are outlined at a generalized and high level here for a "fully aware" vehicle; early systems will provide a subset of these capabilities.

- Situational Awareness: Detection of other vehicles, motorcycles, pedestrians, bicyclists, animals, and any objects of a nature that could be a threat to the vehicle integrity; awareness of speed limits; awareness of traffic control devices and their meaning (stop signs, traffic signals, turn restrictions, lane restrictions, one way streets, and so on); awareness of signaling from other vehicles (turn signals); detection of road features (lane boundaries, entrance/ exit ramps, and so on); detection of traffic circles and configuration, awareness of vehicles already in the traffic circle; awareness of right of way for other vehicles; detection of road condition; exchange of operating data with nearby vehicles; understanding work zone configurations sufficient for safe and lawful maneuvering.
- Operating Environment: Operation in a wide variety of weather conditions and all levels of traffic density.
- Behavior: Exhibiting appropriate behavior in lane changes and at merge points, intersections, and traffic circles; observing traffic etiquette.
- Driver Vehicle Interaction: Ability to assess driver's readiness to resume control (Levels 2 and 3); ability to place vehicle in a safe state if driver is not responding to a request to take over control; provision of appropriate feedback to driver as to system state.

Intervehicle coordination is a special case. For cooperative headway management, intervehicle communications are used to implement cooperative adaptive cruise control (C-ACC). Using V2V communications, a lead vehicle's braking and other control data can be received instantaneously by a vehicle following immediately behind under automatic longitudinal control. Compared to assessing deceleration based on radar sensing alone, the wireless data provides information more rapidly to the following vehicle if the lead vehicle is braking, so that the following vehicle can brake appropriately to maintain the desired headway and/or avoid a crash. C-ACC is essentially a two-vehicle platoon. For automated platooning of several vehicles at close headway, intervehicle communications are used to provide information on lead vehicle actuation to all following automated vehicles. The tasks of the following vehicles are limited to following the vehicle ahead, maintaining the headway, and join/split maneuvers.

2.7 Industry Activity

Several forms of automated vehicle operation are being pursued by the industry. This chapter outlines industry activity in each of these areas, for both passenger car and truck platforms, and for both autonomous and cooperative automation. First, various forms of automated highway driving are examined, then city street driving. Automated parking and operation on unimproved roads are discussed as well, given the potential synergy with on-road applications.

2.7.1 Highway Driving

2.7.1.1 Level 1 Systems

C-ACC is a Level 1 system enabling shorter following distances, which translates into significant improvement in fuel economy (via improved intervehicle aerodynamics) and greater per-lane vehicle flow rates. Because heavy trucks benefit the most from aerodynamic flow improvements, the U.S. trucking industry is beginning to investigate the feasibility of C-ACC for reducing fuel consumption. Within single large trucking fleets, enough trucks are near one another on roadways between freight hubs that these units could coordinate to link up via V2V communications and greatly reduce the headway below that driven now by even adventurous truckers. C-ACC capabilities could be retrofitted to existing heavy trucks, but first technical integration, reliability, user acceptance, regulation, and business case development must be addressed. Truck manufacturers and fleets are expected to begin investigations of CACC under USDOT funding beginning in 2013 [2-38].

For passenger cars, C-ACC user acceptance research has shown that drivers quickly become comfortable with smaller intervehicle gaps than those offered by today's ACC systems [2-39]. C-ACC experiments in Europe have demonstrated traffic flow and flow stability benefits [2-40]. With C-ACC offering both personal and societal benefits, USDOT is expected to begin working with industry to further investigate this opportunity in the near term. Product introduction depends on sufficient penetration of V2V-equipped vehicles for customers to achieve value.

2.7.1.2 Level 2 Systems

2.7.1.2.1 Low Speed

Traffic jam assist (TJA) provides full control of driving in congested conditions and functions on limited-access highways at slow speeds. Building on adaptive cruise control, TJA maintains a safe distance between vehicles, automatically controls the speed and steering, and is able to stop the car if necessary. BMW customer acceptance studies found that two-thirds of respondents felt positive about having traffic jam assistance available on future cars [2-41].

No one likes to drive in a traffic jam, so TJA is expected to move into the market quickly. Mercedes introduced TJA as an option for Model Year 2014 [2-42]. The Mercedes system employs a sensor suite enhanced with stereo vision, which enables normal lane marker detection as well as car-following when in a traffic jam. The driver is expected to be engaged, with hands on the steering wheel; if this is not the case, the system will disable after a few seconds.

Not to be outdone, BMW Group now offers TJA on their MY14 5 Series vehicle. This is a fairly high volume vehicle and as such will greatly increase the market momentum for automated driving. BMW had previously implemented a TJA system in their i3 concept vehicle, which operated at speeds up to 25 mph (40 km/h); the system required the driver to keep one hand on the steering wheel [2-43].

Other companies are demonstrating TJA systems and setting the stage for near-term introduction. Audi has demonstrated "piloted driving" on a Nevada freeway in public traffic

[2-20, 2-44]. The system provides automated control of steering, braking, and throttle at speeds up to 37 mph (60 km/h). Sensing is based on radar, a wide-angle video camera, a laser scanner, and eight ultrasonic sensors monitoring zones directly in front of the car and at the corners. The perception sensors enable the vehicle to define a de facto lane in the absence of lane markings, and the system is capable of detecting objects such as pedestrians and guardrails in addition to other vehicles. The system also responds cooperatively to cars moving into or out of the lane. If the vehicle reaches the limits of the TJA function–such as dispersal of the traffic jam–the driver is prompted to take over control. Volvo Cars has announced plans to launch TJA in 2014 for initial use in heavy traffic at speeds up to 31 mph (50 km/h) [2-17]. Ford is developing a TJA system for the 2017 timeframe, which would be able to communicate developments regarding surrounding traffic to the driver and incorporate features to help the driver remain alert and in contact with the vehicle controls [2-19].

2.7.1.2.2 Highway Speed

Several car companies are focusing on lane-keeping assistance at normal highway speeds with varying degrees of steering support. When combined with ACC, driving can become a form of steady-state Level 2 automation. In Europe, the Honda Accord is available with a Lane Keeping Assist System (LKAS), which provides up to 80% of steering force required [2-45]. The system is designed to complement, not replace, the driver; a warning is provided if the system detects that the driver has taken their hands off the steering wheel. In addition, subtle lane-keeping assistance is being developed that counteracts steering disturbances such as ruts in the lane and crosswinds; Nissan has introduced this capability in a Model Year 2014 vehicle. Called Direct Adaptive Steering, Nissan states that fewer fine-grained steering adjustments result in reduced driver fatigue and workload [2-46].

General Motors has implemented combined ACC and lane keeping in their "Super Cruise" prototype, which has been tested extensively with regular drivers on a test track. GM's system allows drivers to take their hands off the steering wheel for extended periods; however, the driver is still in charge of the vehicle. Therefore, knowledge of the driver state and vehicle capability is essential, and GM has invested significant effort into understanding transfer of control. Countermeasures have been developed in case the driver's attention to driving is insufficient. Super Cruise is intended for the Cadillac brand by mid-decade [2-47].

BMW has demonstrated a Level 2 vehicle combining ACC and lane keeping. When slower traffic ahead prevents the vehicle from maintaining its set speed, the vehicle monitors the adjacent lane and performs a lane change when safe. The vehicle also adjusts speed to allow merging vehicles to enter traffic safely. The system has driven the A9 motorway between Munich and Nuremberg with no driver intervention, as part of over 6214 miles (10,000 km) of testing. During 2013–2014, BMW is working with Continental to equip several near-production Level 2 vehicles for pilot testing on European motorways, taking into account motorway intersections, toll stations, work zones, national borders, and other challenging situations (Figure 2.6) [2-43, 2-21] (BMW Shows Off Their Semi-Autonomous Driving System: http://www.youtube.com/watch?v=9V-vfzzfhzI).

Figure 2.6 BMW's highway driving prototype during testing. (courtesy of BMW Group)

Within the European HAVE-it project, the Automated Assistance in Roadwork and Congestion demonstrator was developed [2-48]. The aim was to support drivers with ACC and lane keeping while driving through a work zone, including the possibility that lane lines may not be visible or accurate in the work zone; therefore, additional objects such as adjacent-lane trucks, beacons, and guide walls were used for guidance. A control algorithm called "virtual wall" steered the vehicle back into the lane when it got too close to a conventional or unconventional lane border.

Initial versions of Google's much-heralded automated vehicle can be considered a Level 2 system. The vehicle drives in automated mode, staying in its lane (via a virtual mapped environment) and braking for slower moving or stopped traffic ahead (via on-board sensors). A user interface displays system state information, notifies the driver when automated mode is available, and prompts the driver to retake control when needed. Press reports note that Google has logged 500,000 miles (804,672 km) of automated driving [2-37]. Latest results show 96,000 miles (154,497 km) between critical safety interventions, and there have been no crashes under computer control (Self Driving Car Test–Steve Mahan: http://www.youtube.com/watch?v=cdgQpa1pUUE).

2.7.1.3 Level 3 Systems

Some of the Level 2 systems described could potentially operate as Level 3 systems under certain conditions and within certain environments. In recent work, systems designed for Level 3 operation have generally been in the platooning domain.

Passenger car platooning was first implemented in USDOT's AHS program in the 1990s, and similar work took place in Japan during that period. Work in truck platooning has also been pursued around the world, as noted in the preceding History section. UC-Berkeley's extended program in truck platooning culminated in 2010 when researchers measured fuel economy savings of 4.3–14.5%, depending on platoon position, running at 6 m gaps at highway speeds [2-3, 2-49]. The Japanese Ministry of Economy, Trade and Industry (METI) began the Energy ITS project in 2008, focusing on automated truck platooning [2-50]. The work is being conducted by a variety of industry and university researchers, led by the Japan Automobile Research Institute. A platoon of three automated trucks driving at 50 mph (80 km/h) with 10 m gaps has been achieved, with further work planned. Technologies include lane marker detection for lateral control via computer vision, plus gap measurement for longitudinal control using radar, laser scanner, and intervehicle communications (NEDO Japan Unmanned Vehicle Technology –Automated Trucks: http://www.youtube.com/watch?v=vIX6u6qHTK4).

The first passenger car platooning work in Europe began in 2009 with the SARTRE project, in which vehicles from Volvo Cars and Volvo Trucks were equipped for close-headway platooning [2-51]. The project investigated the business model of a human lead driver paid by occupants of the fully automated following vehicles in the platoon. Prototypes were developed using current production technology plus V2V communications. During the project, several hundred kilometers of testing was performed on public roads in Spain. At the final project demonstration in 2012, cars and trucks drove automatically behind a leader truck at 85 km/h, separated by 5–15 m (SARTRE Platoon Travel: http://www.youtube.com/watch?v=6YxOKp7lKhs). Some have argued that the SARTRE approach to platooning is a near-term automated driving system, in that it relies on human intelligence in the lead vehicle to handle any anomalies in the traffic stream. Others have countered that asking the public to accept both automation and close-headway operations in a first-generation automation offering may not be realistic.

2.7.1.4 Level 4 Systems

Level 4 systems for the highway setting are expected to emerge from Level 3 systems over time. One special case, however, could apply in the near term: How should the vehicle respond when a driver becomes incapacitated? BMW has prototyped an Emergency Stopping Assistant (ESA), which employs biosensors to detect a medical emergency (such as a heart attack) and takes over driving [2-21]. The system activates the hazard warning lights and maneuvers safely through traffic to come to a stop on the road shoulder. In addition, an emergency call is automatically sent out to request medical assistance and to notify the traffic authorities. European researchers are expected to conduct further work on this topic, and a similar system is being developed within the Japanese Advanced Safety Vehicle program [2-52].

2.7.2 City Street Driving

2.7.2.1 Level 2 Systems

The University of Braunschweig in Germany entered their Leonie vehicle into the DARPA Urban Challenge and has continued to develop its capabilities since then (Figure 2.7) [2-53]. This research focuses on automated urban driving, with testing conducted in live traffic on the city's "Stadtring." This is a busy four-plus lane arterial with traffic lights, turn lanes, pedestrians, and so on. The vehicle executes turn maneuvers, responds to traffic lights, and performs lane changes as needed to flow in traffic (Driverless Car: "Leonie": http://www.spiegel.de/video/fahrerloses-auto-leonie-kreuzt-durch-braunschweig-video-1088474.html).

Figure 2.7 The University of Braunschweig Leonie vehicle.

The Automation Basse Vitesse project, being pursued by IFSTTAR (Institut Français des sciences et Technologies des Transports de l'Aménagement et des Réseaux) in France along with several industry partners, is implementing automated operation in congested traffic in urban areas, both arterials and freeways [2-54]. Two demonstrator vehicles are being constructed that provide full automation up to 31 mph (50 km/h). The technical focus is on perception and data fusion, and traffic flow and legal impacts are also being assessed.

2.7.2.2 Level 4 Systems

Cybercars are specially developed Level 4 automated urban vehicles for operation in city centers in lanes or districts segregated from normal traffic. Development has been ongoing in Europe since 2000, including demonstrations and pilot testing in several cities. The CityMobil2 project is aimed at deployment of these systems [2-55]. The project runs from 2012–2016, during which two sets of six vehicles will perform five city demonstrations.

GM has developed an approach to personal urban mobility with their EN-V vehicle, unveiled in 2010 (Figure 2.8) [2-56]. The GM urban mobility vision calls for vehicles connected to each other, other road users, and the road; electric propulsion; a small physical footprint; and dedicated zones for automated operations enabled by the cities. Within this vision, GM sees automated driving improving the business model for mobility-on-demand systems. An EN-V2.0 pilot deployment is under development in Tianjin Eco-City, China.

Figure 2.8 GM's second generation EN-V urban vehicle. (courtesy of GM)

2.7.3 Automated Valet Parking

Automated valet parking is a Level 4 system that will be a useful stepping stone to on-road automation. Audi has developed a prototype system that enables the driver to leave the vehicle at the entrance to a parking garage and instruct it to park itself via a smartphone interface [2-20]. The vehicle returns when summoned by the owner. Audi is currently equipping a parking garage in Germany for further system testing (Video: CES in 60: Audi Piloted Parking http://abcnews.go.com/Technology/video ces-60-audi-pilotless-parking-18165480).

Nissan's self-parking car system functions similarly, with another layer added: Upon receiving the park request, data is sent to the Nissan Global Data Center, which performs a vehicle health check before enabling the automatic driving mode, providing an extra level of confidence in the vehicle's proper operation [2-57].

2.7.4 Automated Operation on Unimproved Roads

A diverse set of players are developing Level 2-4 automated vehicle capability for all road operation (including unpaved roads and difficult terrain). These are briefly summarized here, as this technology could be applied to on-road production systems in the future.

The U.S. Army has developed the Convoy Active Safety Technology system [2-58], which is intended as a low-cost automated following system for tactical wheeled vehicles (CAST Makes Autonomous Convoys a Reality: http://www.youtube.com/watch?v=CK5cv8ZzJNI&list=PL0D2B4BF9398B639C&index=5). The system maintains an intervehicle gap, adjusting vehicle speed, shifting gears, and handling obstacles as needed. Following vehicles can assume command of the convoy automatically if the lead vehicle becomes inoperable. The program achieved more than 100 miles of continuous automated driving with a 50 mph (100 km/h) capability on paved roads and 35 mph (56 km/h) capability on dirt roads. The army is moving a step closer to full production with the Autonomous Mobility Appliqué System, designed to retrofit existing military trucks with a range of commercial automotive systems, from active safety to full automation [2-59]. Additionally, the U.S. Office of Naval Research is pursuing the Small Unit Mobility Enhancement Technology (SUMET) program, aimed at traversing difficult terrain supporting U.S. Marine Corps operations such as overwatch surveillance [2-60]. SUMET is pushing the envelope of autonomous automated capability achievable with low-cost camera sensors only–thus a completely passive sensing platform. The system can classify terrain and material (wood, foliage, dirt), as well as typical objects, to find its way through unknown territory with only an end point defined by an operator.

Similarly, Israel's Ministry of Defense is now operating unmanned automated vehicles for border patrol [2-61]. These vehicles have taken the place of some soldiers on the front lines, preventing them from facing possible hostile activity.

The mining industry has been a pioneer in adopting automation due to gains in safety and productivity. The Rio Tinto Yandicoogina mine in Western Australia is now operating a fleet of ten Komatsu driverless haul trucks, with plans to eventually deploy 150 of these vehicles [2-62].

2.7.5 Commentary

This broad array of activities, led by the auto industry, is a strong indication that road transportation is turning the corner toward co-pilot and autopilot functions. Full automation will focus initially on low-speed congested highway traffic situations–clearly the most tedious of driving experiences, and therefore car customers will see high utility in purchasing such a system. The speed allowances on these systems will gradually increase to cover the full range of highway speeds. Trucks will begin dual-vehicle close-headway operations in the near-mid-term; platooning for passenger cars is most likely a later generation capability. At this time, automation in urban settings has a limited degree of industry investment; however, specialized applications are very promising. Off-road systems have their own market momentum, which may result in synergy with on-road operations.

2.8 Government Stance

Governments are active around the world in exploring vehicle automation. This section provides a short overview.

2.8.1 Asia

The Japanese Ministry of Land, Infrastructure, Tourism, and Transport (MLIT) conducted research in vehicle automation during the 1990s and is now planning a new program [2-63]. Technical and safety-related investigations are ongoing, and institutional, social acceptance, infrastructure, and socioeconomic impacts are being studied as well. Realization of "autopilot" is seen in the early 2020s, with V2X communication providing traffic information, detailed road maps, and other data. Currently, a Level 1 pilot project is under way to counter congestion at "sag sections" of the roadway (a slight downhill followed by a moderate incline in which vehicles tend to slow down without the driver's knowledge). The sag congestion countermeasures rely on the central traffic management center broadcasting to C-ACC–equipped vehicles in specific road segments to maintain appropriate intervehicle distance and thus maintain traffic flow. In parallel, MLIT is pursuing the Advanced Safety Vehicle–Phase 5 Program, which is focusing on advanced crash avoidance. The Ministry of Economy, Trade and Industry (METI) is sponsoring the Energy ITS project focusing on automated truck platooning, with a key aim to reduce emissions.

Other work in Asia is occurring at universities, supported by government and industry sponsors. In Korea, Hyundai-Kia sponsors an Autonomous Vehicle Competition annually, funding selected university teams. Recent news reports from China described the Fierce Lion 3 automated vehicle, which was developed by the Military Transportation University of the PLA [2-64]. Testing of 10,000 km has been completed, performing automated cruising, lane changes, and passing of slower traffic.

2.8.2 Europe

The European Commission (EC) has invested heavily in advanced driver assistance systems in recent decades, and vehicle automation is now a priority [2-65]. Officials have noted that "automation is increasingly seen as the only long term option." Automation is seen as providing drivers with assistance in complex driving situations and taking over some driving tasks, while on a societal level offering great potential to improve safety and energy efficiency. In recent years, the EC joined with industry partners to fund the HAVEit project investigating "highly automated driving"–essentially a Level 2 system combining ACC and lane keeping– plus the SARTRE project investigating business models for close-headway platooning. New projects started in 2013 focus on "fault-tolerant and resilient supervised automated driving" and the potential use of "smart lanes or dedicated lanes for automated vehicles." Specific areas of interest include driver take-over situations, automated emergency stopping, and socioeconomic, standardization, and legal issues.

On the national level, the German UR:BAN project is funded by the German Federal Ministry of Economics and Technology and cost-shared with the vehicle industry [2-66]. The research is aimed at crash avoidance in urban environments, energy efficient driving, and anticipatory driving in urban scenarios. The project will deploy an intelligent infrastructure to communicate with vehicles for optimizing traffic efficiency. Legal issues will also be examined.

France has been the center of Cybercars work at the French National Computer Science Research Institute, and the Automation Basse Vitesse project is being conducted under IFSTTAR leadership, as described earlier [2-54]. VeDeCoM, a new Institute of Excellence, is bringing together local authorities and players in the automotive industry to conduct joint research, training, and incubate new start-ups. New vehicle systems and connectivity (including cooperative systems, driving assistance, automation, and platooning) are a major focus, as well as research on technology and socioeconomic questions for innovative mobility systems.

2.8.3 United States

As product development in vehicle automation ramps up in the private sector and the states allow public road testing, USDOT is turning its attention to assessing the risks, addressing the challenges, and developing opportunities. In parallel, USDOD is beginning trials for specific vehicles and operational environments.

Within the USDOT Intelligent Transportation Systems (ITS) program, the ITS Joint Program Office (JPO) plays a strategic role, with the Federal Highway Administration (FHWA) responsible for road and traffic aspects and the National Highway Traffic Safety Administration (NHTSA) responsible for vehicle safety. The rise of an automated driving focus comes as their Connected Vehicle program is reaching a decision point concerning a possible mandate of V2V communications on all new passenger cars. Maintaining that the full potential benefits of vehicle automation can only be achieved through a connected environment, the ITS JPO expects to complete a Vehicle Automation Program Plan by in 2014 [2-36]. FHWA is planning research into C-ACC for both passenger cars and heavy trucks, a precursor to connected automation. In 2012, NHTSA announced a comprehensive Vehicle Automation Research Program, which addressed driver role, human factors, performance metrics, and other topics [2-67]. NHTSA officials have said it is not enough for an automated vehicle to be "as safe" as human drivers–it has to be better than a human–as well as "crash-less." Their Motor Vehicle Automation Roadmap is aimed at supporting policy decisions on emerging system concepts, developing technical requirements and performance tests, and assessing safety benefits. A study was launched in 2012 to examine human factors issues for automated driving Levels 2 and 3. The research will address the degree to which drivers can safely interact with equipped vehicles, including effective control handoff strategies and human-machine interface concepts.

The U.S. Department of Defense (USDOD) began the Autonomous Mobility Appliqué System (AMAS) project in 2012, aimed at developing kits to retrofit existing army trucks for automation, convoying, and crash avoidance [2-59]. Other DOD work in addition to that described previously focuses on using military facilities as test beds for automated operations; for instance, a project under development at Fort Bragg would use automated vehicles to transport wounded soldiers from barracks to medical facilities on the base [2-15].

2.8.4 Collaborative Efforts

Significant industry-led, scientific, and international collaborative efforts focused on automation have recently arisen. In Europe, the Automation Working Group within the iMobility Forum, formed in 2012, is addressing automated driving issues, including technical, financial, organizational, and legal issues [2-68]. In 2013, several additional collaborations were launched. The European Commission, Japanese MLIT, and USDOT began a Tri-Lateral Working Group on Vehicle Automation R&D; the Transportation Research Board, part of the U.S. National Academy of Sciences, formed a joint subcommittee on Challenges and Opportunities for Road Vehicle Automation; and the Fleet Automation Forum was formed in the United States as a user-centric forum to develop working groups and define pilot testing critical to adoption of automated driving by the trucking industry [2-38].

2.9 Engineering and Human Factors Considerations

The technologies to accomplish automated driving fall into the categories of situational awareness, decision-making, and actuation. Situational awareness relies upon sensors, maps, positioning, sensor fusion, and software. Decision making is software based, requiring highly sophisticated algorithms at the planning and control levels. Actuation is well advanced due to the advent of driver assistance systems. Reliability is a cross-cutting regime that falls well within automotive best practices, yet automation will present some new territory that must be addressed. Similarly, cybersecurity is a critical yet "here and now" issue due to the current proliferation of vehicle connectivity channels that could potentially be used to activate electronic vehicle controls; it is not yet clear whether automation will present additional cybersecurity challenges.

2.9.1 Situational Awareness

Development of active safety systems for collision avoidance has progressed sufficiently that current systems have reasonably high situational awareness of nearby vehicles, pedestrians, bicyclists, and road elements, but atypical situations are left to the human driver. For automated operations, one of the most vexing perception challenges is in interpreting signs and signals at temporary road construction zones, which in some instances can be chaotic and challenging even for human drivers [2-69]. In the United States in particular, signage and markings in these areas vary widely from state to state or from one work zone to the next. Cooperation with state and city governments to bring some level of standardization to work zones would be very helpful.

2.9.2 Automated Vehicle Behavior in Traffic

Behaviorally, protocols need to be clear for actions taken when the vehicle cannot handle the situation. In the HAVEit project, a minimum risk maneuver was implemented for this situation: the vehicle either performed hard braking to a stop or changed lanes and safely brought the car to a stop on the shoulder (essentially the emergency stop assistant described previously) [2-45]. When a driver takeover request is issued by Audi's prototype TJA, it addresses a lack of driver response by tapping the brakes after 5 seconds; if there is still no response, the car begins to gradually brake to a stop within the next 10 seconds and activates the hazard flashers [2-20, 2-44]. Although these actions are not ideal, they are preferable to a vehicle driving "blind." Since this vehicle behavior impinges on other road users it raises the question, is consistency needed across the car industry so that other motorists are not confused? More generally, what will it take for society to accept machine decision making in traffic? The key could be in the ability to learn; possibly machine replacement of human drivers will come when the system is able to be aware of its mistakes and not repeat them.

An even higher level of performance could be a system that is designed to learn from and avoid the mistakes of other automated systems on the road, via vehicle connectivity [2-70].

2.9.3 Test Methods

Extensive test protocols are used by the auto industry to certify that advanced driver assist systems are ready for market (Figure 2.9) [2-33]. While useful, some industry experts believe new and specific development and test procedures are needed for testing automated driving [2-71]. Developers have reported many miles of testing, but miles alone are not the issue–the performance of the system under a variety of scenarios encountered during high-mileage testing is the key, with test protocols designed to expose the system to as many challenging and unpredictable scenarios as possible. Testing must also address areas such as functional robustness, software safety, and human-machine interaction. Regulators will likely play a major role in defining performance requirements and test and evaluation methodologies, particularly in terms of objective testing procedures [2-67].

Figure 2.9 Driving simulator testing within a SARTRE platoon. (courtesy of Volvo Car Corporation)

2.9.4 Exceeding Human Performance

In terms of human capabilities, the industry needs a baseline for human performance in driving. When goals such as "better than human driving" are stated for vehicle automation, what does that mean? Potentially, the vehicle could deliver the same overall performance as a "perfect" driver in momentary steady-state driving while also delivering statistically better performance than the "average" driver over the longer term [2-41]. Metrics are needed for both normal driving and crash-imminent situations.

What is the role of the driver? A range of industry positions have been stated. Rather than a car that drives itself, Toyota is focusing on technologies that support "an intelligent, always-attentive co-pilot whose skills contribute to safer driving" [2-72]. Ford aims to "establish a rapport between the car and the driver" and not remove the driver completely from the process [2-73]. GM sees the driver as relaxed but engaged and aware, still the "captain" of the vehicle and in control [2-74, 2-75]. The theme continues, with BMW holding that the driver must stay aware of the surrounding conditions and "remain responsible for the situation at all times" [2-43]. Japan's MLIT holds a similar view; their Advanced Safety Vehicle research team has adopted a design philosophy stating that drivers play the main role in driving safely [2-52].

However, others are pointing to the customer's core desire–to be relieved of the tedium of driving. Audi notes their TJA system "enables the driver to devote their attention, within certain limits, to other activities while the system is operating" [2-20]. Volvo Cars sounds a similar note, asserting that automated driving paves the way for "more freedom behind the wheel" [2-17, 2-18].

2.9.5 Driver Monitoring

Driver monitoring may play a key role. Driver fatigue systems now on the market monitor driver activity by noting lane keeping stability, pedal application, steering inputs, manipulation of other controls, and direct driver monitoring. Toyota's driver monitoring system uses a camera to track the movement of the driver's head as an indicator of level of attention towards the road [2-72]. Also, for LKA systems offered by several carmakers, hands-on wheel detection is accomplished through counter-torque detection or capacitive touch sensing. Some argue that driver monitoring is essential for Level 2 automated driving in particular, to better understand the driver's attention to the road situation and ability to resume control if needed. However, as noted by HAVEit researchers, automated driving reduces or eliminates data elements now used for driver monitoring [2-48]. They concluded that vision-based measures, which directly assess the driver's state by analyzing the driver's eye closure behavior for drowsiness detection and the driver's head or gaze direction to detect distraction, are the most promising.

Ford's driver workload estimator uses data from active safety sensors to continually assess the intensity of the traffic situation around the subject vehicle (Figure 2.10) [2-76]. Biometric data from the steering wheel provides the driver's heart rate and body temperature. Infrared sensors on the steering wheel monitor the palms of a driver's hands as well as his or her face, looking for changes in temperature. Additionally, a sensor embedded in the seat belt assesses the driver's breathing rate. With greater knowledge of the driver's state, vehicle systems can dynamically adjust alerts and filter interruptions. These types of techniques may translate to monitoring for vehicle automation.

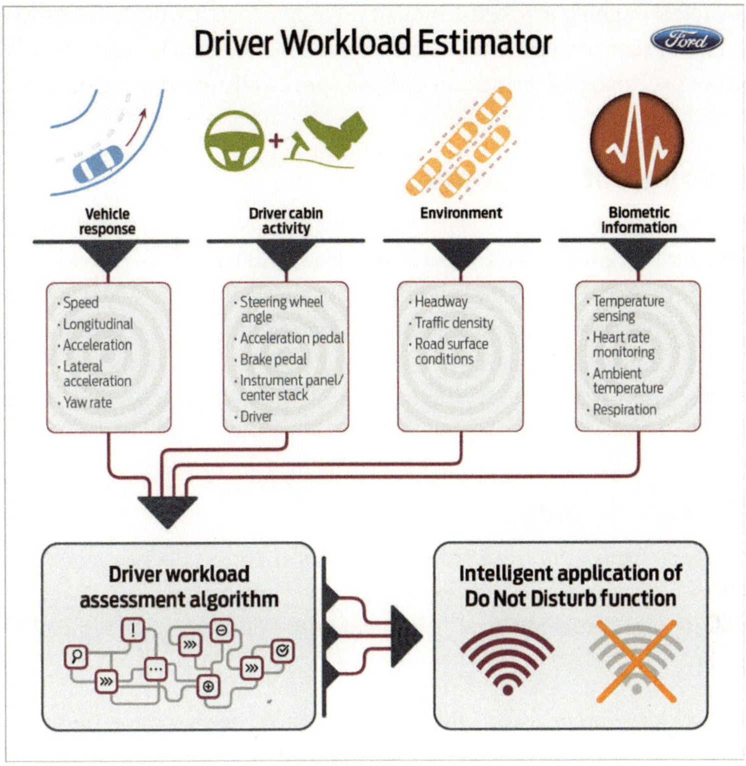

Figure 2.10 Ford's approach to estimating driver workload. (courtesy of Ford Motor Co.)

2.9.6 Effective Driver-Vehicle Interaction

The driver-vehicle interface is critical in an automated driving system. The method of engaging and disengaging control will be key, and well as techniques to maintain the appropriate level of engagement. How much advance warning time is enough for the driver to safely engage? The driver engagement time could vary by the driver type–novice, experienced, impaired. The answer could also depend on specific driver task requirements and use of driver monitoring. Ideally, when the system needs to return control to the human, the driver sees it coming, the context is understood, and it will be a mutual decision [2-70]. Given rapidly changing demographics worldwide, special attention should be given to requirements for older drivers in automated driving.

The German UR:BAN project adapts driver support systems to the driver's current state, whether overburdened or inattentive [2-77, 2-66]. Ideally, the driver receives the right information to anticipate traffic situations rather than simply to react to them. Going one step further, UR:BAN researchers are developing methods of identifying the intentions of the driver of the subject vehicle plus other drivers as early as possible, so that this information can be taken into account in the subject vehicle's response.

In-depth research is ongoing across the industry. For instance, Toyota is working with Stanford University to prototype and evaluate an automated vehicle driver-vehicle interface based on a set of psychological principles defined specifically for automated driving [2-78].

2.10 Deployment Challenges

The key deployment issues going forward are primarily institutional and regulatory, rather than technical. For Europe and the United States, legal and liability issues dominate. Like any technological system, there is the possibility for a system failure with automated vehicles, and the consequences could be tragic. Extensive tools and procedures developed by the auto industry for evaluating active safety systems reduce the likelihood of catastrophic failure but cannot eliminate the possibility entirely. Thus the legal system will have to grapple with automated vehicles in some manner.

2.10.1 U.S. Legal Situation

The Stanford Center for Internet and Society (CIS) has analyzed the key legal issues regarding automated vehicle sale and operation in the United States [2-79]. Simply put, automated vehicles are "probably legal." However issues do exist in the regulatory and liability domain.

In the United States, the premise for vehicle regulations is "everything is permitted unless prohibited." Three legal regimes define the playing field: the 1949 Geneva Convention on Road Traffic, regulations enacted by NHTSA, and the vehicle codes of all 50 states. CIS concludes that the Geneva Convention "probably does not prohibit automated driving." The treaty requires every vehicle to have a driver who is "at all times ... able to control" it, which could be fulfilled by the driver's ability to override the automated system. In addition, no current NHTSA regulations prohibit or impede the sale of automated vehicles.

Nevada, Florida, and California are currently the only states to expressly regulate "autonomous vehicles," although legislative activity continues in other states. Rather than "legalizing" automated vehicles, they serve to endorse the potential of automated technology and take steps such as defining terms, establishing requirements for R&D testing, and establishing basic conditions for legal operation. In particular, California is developing a regulation by January 2015 to certify automated vehicles for public road use. This has galvanized the auto industry to develop recommendations as to how this might be accomplished, as this law could lay the groundwork for national level requirements. On a separate note, following distance requirements will almost certainly need to be adapted to support C-ACC and platooning at some point in the future.

In terms of liability when automated vehicles are involved in crashes, some studies have concluded that manufacturers' product liability could increase [2-80]. To the degree this proves to be the case, manufacturers could be inhibited from bringing such systems to market.

2.10.2 European Legal Situation

Some legal experts in Europe have concluded that full automation would require a new approach to the legal framework in road traffic. For higher levels of automation, similar to the U.S. situation, manufacturers could be held liable in any incidents in which automated control was involved, based on current law [2-81].

The Convention on Road Traffic (Vienna 1968) holds sway over the European market for automated vehicles. Language addressing driving requirements–that the vehicle has a driver, that the driver is able to control the vehicle, and that the driver is attentive–impinges directly on automated driving. In particular, Article 5 states "every driver shall at all times be able to control his vehicle or to guide his animals." Does this prohibit automation, or can terms like "control" be reinterpreted as the "power to take control, power to monitor" and in such cases the driver stays "in control"? Some see such an interpretation as the way forward, while others are arguing for amending the Convention with new language–which could take significant time. Among others, Verband der Automobilindustrie, the German automakers association, is working to address such issues [2-71]. All indications are that much more debate lies ahead, even while initial automated systems such as TJA are being rolled out.

For the personal urban vehicles being pursued in CityMobil2, unique challenges exist due to unclear, non-existent, or incompatible legal frameworks [2-82]. In the safety and certification domain, an established criteria and method to certify these unique systems as "safe enough" is needed. To get permission from local authorities for recent system demonstrations, certification procedures were developed that were acceptable in some cases. However, the Abu Dhabi Ministry of Transport required an additional analysis before the system was allowed. The overall situation creates a costly, time-consuming, and unpredictable process. The current project is drafting formal procedures that could be adopted across Europe as a future certification standard for such systems [2-55].

2.10.3 Market Dynamics

As noted previously, the liability climate could inhibit manufacturers from bringing automated vehicles to market. However, it must be understood that vehicle manufacturers live with lawsuits as a business reality; at a practical level, their aim is not to avoid lawsuits altogether but to contain risk. The degree of risk is balanced by competitive pressures, that is, the "feature wars" that keep innovations of all sorts coming to new cars. These calculations are difficult yet necessary and have been made in the introduction of every collision avoidance product introduced in the last 15 years.

Observers of activity to develop vehicle automation products can raise numerous questions about "what can go wrong"; however, it should be remembered that the market dimension trumps a complete technical solution. Although vehicle manufacturers will proceed with great care, not all questions need to be answered for market introduction; as a case in point, telematics came to market without the need to "prove" that the systems did not create driver distraction. For telematics, as well as with automation, personal responsibility of the driver to attend to the driving task is fundamental. The carmakers have high standards for testing, and they will rely on their strict internal criteria to bring systems to market.

2.11 Looking Down the Road

2.11.1 Considerations

To bring automated driving systems to market, product developers must make decisions across a multidimensional space. One dimension is the driving function–driving on highways only or all roads; driving only in certain situations (traffic jams) or in all situations; operating in a single lane or automatically completing lane changes. Across this space is mapped the driver role. For any of these operational scenarios, the driver role could be defined as close monitoring, relaxed monitoring, or no monitoring (aligning with automation levels 2, 3, and 4). Though automation levels are well defined, actual use could be a mix; for instance, steady-state highway driving could be done with relaxed monitoring, and the vehicle could request confirmation from the driver for a lane change, which the driver would monitor closely for the few seconds that this would take.

A few key variables will likely define the near-term evolution of automated vehicles for highway driving–the domain where the vehicle industry is now investing heavily. These are speed (low/high), staying in lane or changing lanes, and degree of driver monitoring (if any). What about mapping? Some approaches rely on the roadway being pre-mapped with sensor-relevant data. One concept calls for the vehicle to learn routes as the owner drives manually. For frequently traveled routes (such as home to work), the vehicle would quickly gather the necessary data and "turn on" that route for automation. Or, it may operate with a differing level of driver monitoring on pre-mapped versus unknown routes. Google relies strongly on a preexisting database, and the automotive industry is beginning to define an approach to map supported systems as well [2-70].

Product decisions about driver monitoring raise interesting questions in regard to personal responsibility. A carmaker could assert that the system operation and driver responsibilities are clearly defined in the owner's manual, plus require an "I accept" button on the vehicle screen to acknowledge a message reinforcing the driver's responsibility (this was done when integrated navigation systems were first introduced). Under such a philosophy, it is up to the driver to fulfill their responsibility, and no driver monitoring needs to be implemented. The opposite viewpoint is to use driver monitoring to make sure the driver is paying attention properly.

In this context, let us examine a parallel example. Currently, drivers control the speed of their vehicle. Should a vehicle system monitor the driver's chosen speed, determine if it is inappropriate for conditions, and potentially intervene to slow the vehicle? A theoretical top-down safety engineering approach may call for this; however, the driving regime has not evolved in this way. The choice of speed is left to the driver, even if this endangers other road users. Countermeasures for unsafe behavior are in the domain of law enforcement. So, for an automated vehicle, should the carmaker install driver monitoring equipment to "baby-sit" in the event that a driver chooses not to properly pay attention? Maybe; maybe not. Given the potential for manufacturer liability for incidents occurring under vehicle automation, a new reason for driver monitoring could emerge as well–to protect the manufacturer by gathering data to determine if the driver is using the system correctly. And since the Emergency Stop function applies to a variety of situations beyond the rare event of a driver being physically incapable, the same technology could be employed as a backstop for the driver who tunes out. Conceivably, regulators might require this as a way to protect other road users from irresponsible automated car users. Final product decisions will incorporate these plus a range of other considerations.

2.11.2 Rollout

When will various levels of vehicle automation become available? TJA is here now. Beyond that, there are a variety of viewpoints. Some see very rapid rollout: Google plans to release the technology in some form within the next five years [2-83]. Within the auto industry, the players generally align in their expectations. Further introductions of TJA are expected very soon, followed by a closely monitored Level 2 highway system by mid-decade. Systems will evolve to offer expanded functions (such as lane changes, merging) and require gradually less monitoring. Starting late this decade, we will likely see the advent of automation that provides the driver with "some time" to resume control–this will be an important competitive discriminator. Do I want to buy the system that requires me to reengage in 3 seconds, or the one that can handle situations well enough to give me 30 seconds? The control transition time will steadily elongate between now and then, in essence moving along the continuum from Level 2 to Level 3 to Level 4.

Motivated by strong benefits to the business bottom line, large truck fleets could implement C-ACC during this decade, operating in-fleet vehicles between freight hubs. The market forces for full automation of trucks are not as strong, since adding lateral control does not have the return-on-investment provided by tighter longitudinal contro–at least not until trucks can operate without a driver at all.

As to platooning, the Ford Blueprint for Mobility, as well as forums in Europe, see these coming in the next decade [2-74, 2-84]. In the urban driving domain, TJA will likely evolve to support start-up at traffic signals and simple arterial road situations with successive traffic lights.

2.11.3 Infrastructure and the Public Sector Role

What should the government do (or not do)? Since industry activity is focused on mixed traffic, on roads as-is, the public sector role is not about infrastructure technology or concrete for initial automated vehicle operations as it was in the 1990s. However the public sector has an opportunity to take advantage of the capabilities of automated vehicles to improve traffic flow using V2X techniques.

For transportation engineers, the combination of connectivity and control is powerful. Sophisticated traffic management centers of the future could combine surveillance data with predictive modeling to generate optimum traffic speeds for lane-specific micro-sectors of roadways. Called "speed harmonization," broadcasting these advisories to connected vehicles on the road would enable (voluntary) automated speed control for these vehicles, at levels of precision beyond what regular drivers could consistently do. These fine-grained speed adjustments hold the promise of smoothing traffic and possibly preventing severe congestion. Even in times of high demand when some delay is unavoidable, the traffic could be managed optimally. Simulations indicate that not all vehicles would need to be equipped; a minority of controlled-connected vehicles scattered through the traffic stream would be sufficient to have a positive effect. Advanced applications, such as speed harmonization, will track with uptake rates for V2X and automated vehicles but could emerge next decade. However, this depends strongly on the level of investment transportation agencies make in developing the capability and implementing systems to communicate to the vehicles. Japan could emerge as the leader in this area, given their centralized and energetic approach to the deployment of new forms of infrastructure capability.

In the next decade, we may see the advent of passenger car C-ACC as the penetration of connected vehicles increases, particularly with incentives offered by road operators such as access to carpool lanes (which would serve to cluster C-ACC equipped vehicles). In terms of public sector research investment, we can expect to see research activities ramp up quickly as the Japanese MLIT defines their automation program, NHTSA begins multiple research projects in automated vehicle safety, FHWA continues to develop automated-assisted traffic flow improvements, and the EC begins new projects in automation. In the urban automation domain, the CityMobil2 project will select cities for their next round of model deployments, while UR:BAN in Germany and ASV-5 in Japan will invest further in situational awareness for urban environments. These activities will have a significant influence on the evolution of automation throughout the remainder of this decade.

2.11.4 Cross-Industry Collaboration

Thus far, carmakers are generally "going it alone" in developing automated vehicle products. Areas in which cross-industry collaboration could be useful are in addressing the legal regime, defining certification criteria and objective testing, establishing common on-road behavior parameters, and in educating the public and lawmakers so that the capabilities and limitations of automated systems are better understood.

2.12 Conclusion

The current situation with automated driving is quite remarkable and could not have been foreseen even just a few years ago. Almost 75 years after automated driving was first conceptualized, TJA, the first version of this technology, is now offered for purchase in personal vehicles for operation on public roads–a milestone whose importance cannot be overstated. Recent demonstrations have attracted extensive media coverage, bringing the concept of automation very much into the public's mind as an enticing new vehicle technology. The media frenzy about automation remains strong and is increasing, and automation has now strongly taken root within the auto industry as a competitive arena. The public image for automation is currently very good, and customer pull is strong due to the personal and everyday benefits offered.

Earning the user's trust is fundamental: When elevators were first introduced, the public valued the function but didn't trust the technology. Riders initially found comfort in elevator operators, and over time, the operators disappeared and the safety of elevators was established. Today, virtually no one has second thoughts about entering an elevator. Trust for automated vehicles needs to be earned as well, so that individuals trust their automated vehicle as much as they trust an elevator.

Automated driving presents significant challenges, particularly in the human factors area. These will not be "solved"–they will be "handled" within the high safety standards of manufacturers and sufficient for the customer, sufficient for liability and brand image, and if needed, sufficient for regulators. Automated vehicle systems will be designed to adapt to limits of humans and technology and still offer value: as is virtually always the case in the car industry, practicality will trump engineering perfection.

In summary, connectivity and automation are two very powerful forces that are poised to enter the marketplace. The evolution and interaction of these forces will be dominant for some time to come and may actually deliver on a future in which crashes and congestion do not so severely undermine our safety and quality of life. Driving is sometimes delightful, but usually it is either neutral or unpleasant; plus, our productivity is limited. Driving is also uniquely a situation in daily life in which we are aware that our lives could suddenly end due to a crash. Congestion and crashes constitute a form of "unnecessary suffering" in today's society, and we are entering a transitional era in which both could become a thing of the past, with personal mobility taking on a completely different nature, one that is safer, more sustainable, more productive, and more enjoyable.

2.13 References

2-1. Accenture. (February 2011). Consumers in US and UK Frustrated with Intelligent Devices That Frequently Crash or Freeze, New Accenture Survey Finds. Retrieved from http://newsroom.accenture.com/article_display.cfm?article_id=5146.

2-2. J.D. Power and Associates. (April 2012). 2012 U.S. Automotive Emerging Technologies Study: Vehicle Owners Show Willingness to Spend on Automotive Infotainment Features. Retrieved from http://www.jdpower.com/content/press-release/gGOwCnW/2012-u-s-automotive-emerging-technologies-study.htm.

2-3. Bishop, R. (2005). Intelligent Vehicle Technologies and Trends. Artech House. ISBN 978-1580539111.

2-4. Transport Research and Innovation Portal. CHAUFFEUR II. Retrieved from http://www.transport-research.info/web/projects/project_details.cfm?ID=15277.

2-5. National Automated Highway System Consortium (NAHSC). Demo '97 video. Available at http://www.youtube.com/watch?v=Dtj9J1A-PMk.

2-6. Thorpe, C., Jochem, T., and Pomerleau, D. (1998). Automated Highways and the Free Agent Demonstration. Robotics Institute, Carnegie Mellon University. Retrieved from http://www.cs.cmu.edu/afs/cs/usr/cet/www/book6.pdf.

2-7. Defense Advanced Research Projects Agency (DARPA) (undated). Urban Challenge. Retrieved from http://archive.darpa.mil/grandchallenge/.

2-8. DARPA Urban Challenge Video, http://archive.darpa.mil/grandchallenge/gallery.asp.

2-9. Vlacic, L., and Parent, M. (2009). Cybercars 2–Final Report. Retrieved from http://cybercars2.paris-rocquencourt.inria.fr/publications/deliverables/cybercars2_summary_report_version8.pdf/view.

2-10. Lenk, C., Haberstroh, M., and Wille, M. (2011). Interaction of Human, Machine, and Environment in Automated Driving Systems. In Transportation Research Record No. 2243 (pp. 138-145). Washington, DC: Transportation Research Board of the National Academies. Cited in Shladover, S. (2012). Literature Review on Recent International Activity in Cooperative Vehicle–Highway Automation Systems, Report FHWA-HRT-13-025. Federal Highway Administration.

2-11. Zlocki, A. (2012). KONVOI and interactIVe: Truck Platooning and Crash Avoidance. Paper presented at the TRB Road Vehicle Automation Workshop, Irvine, CA. Retrieved from http://onlinepubs.trb.org/onlinepubs/conferences/2012/Automation/presentations/Zlocki.pdf.

2-12. The Thinking Car. Available for viewing from http://www.snagfilms.com/films/title/the_thinking_car.

2-13. Robert Bosch GmbH. Bosch Predictive Emergency Braking System. Available at http://www.youtube.com/watch?v=VHR8AeZrlVc.

2-14. National Highway Traffic Safety Administration (May 2013). U.S. Department of Transportation Releases Policy on Automated Vehicle Development. Available at http://www.nhtsa.gov/About+NHTSA/Press+Releases/U.S.+Department+of+Transportation+Releases+Policy+on+Automated+Vehicle+Development.

2-15. Association of Unmanned Vehicle Systems International (AUVSI). (August 2012). Driverless Car Summit 2012: Conference Report. Retrieved from www.auvsi.org.

2-16. Bigman, D. (January 2013). "Driverless Cars Coming to Showrooms by 2020, Says Nissan CEO Carlos Ghosn." Forbes. Retrieved from http://www.forbes.com/sites/danbigman/2013/01/14/driverless-cars-coming-to-showrooms-by-2020-says-nissan-ceo-carlos-ghosn/.

2-17. Duxbury, C., and Stoll, J. (December 3, 2012). Volvo Stakes Its Claim to Driverless Vehicles. The Wall Street Journal. Retrieved from http://online.wsj.com/article/SB10001424127887323751104578147352673892548.html.

2-18. Volvo Car Corporation (September 2012). Volvo Car Corporation Aims for Leadership within Autonomous Driving Technology. Retrieved from https://www.media.volvocars.com/global/en-gb/media/pressreleases/45718.

2-19. Ford Motor Company (June 2012). Ford Develops "Traffic Jam Assist" and New Parking Technology to Help Address Future Mobility Challenges. Retrieved from https://media.ford.com/content/fordmedia/fna/us/en/news/2012/06/26/ford-developing-new-technologies-to-help-ease-traffic--parking-s.html.

2-20. Audi of America (January 2013). Piloted Driving. Retrieved from http://www.audiusanews.com/pressrelease/3304/149/piloted-driving.

2-21. BMW Group (February 2013). Heading for Europe's Motorways in a Highly Automated BMW: BMW Group and Continental Team Up on Next Step Towards Highly Automated Driving. Retrieved from https://www.press.bmwgroup.com/pressclub/p/pcgl/pressDetail.html?title=heading-for-europe-s-motorways-in-a-highly-automated-bmw-bmw-group-and-continental-team-up-on-next&outputChannelId=6&id=T0137270EN&left_menu_item=node__2367.

2-22. Hansen, P. (2011). "Google's Self-Driving Cars Zoom in on the Future." Hansen Report on Automotive Electronics, 24(6), 1-3.

2-23. Silicon Beat (September 2012). Sergey Brin Hopes People Will Be Driving Google Robot Cars in "Several Years." Retrieved from http://www.siliconbeat.com/2012/09/25/sergey-brin-hopes-people-will-be-driving-google-robot-cars-in-several-years/.

2-24. Delphi Automotive. Delphi Intelligent Forward View Camera. Retrieved from http://delphi.com/shared/pdf/ppd/safesec/ifv100.pdf.

2-25. Rodriguez, S. (November 30, 2012). L.A. Auto Show: Animal Detecting Systems Coming as Soon as 2013. Los Angeles Times. Retrieved from http://articles.latimes.com/2012/nov/30/autos la-fi-hy-autos-laas-animal-detection-night-vision-autoliv-20121129.

2-26. Coxworth, B. (May 9, 2011). Continental Forward Braking System to Get Stereo Vision. Gizmag. Retrieved from http://www.gizmag.com/continental-forward-braking-system-stereo-vision/18594/.

2-27. Delphi Automotive. Active Safety: Delphi Electronically Scanning Radar. http://delphi.com/manufacturers/auto/safety/active/electronically-scanning-radar/.

2-28. Robert Bosch GmbH. LRR3: 3rd Generation Long-Range Radar Sensor. Retrieved from http://www.bosch-automotivetechnology.us/media/db_application/downloads/pdf/safety_1/en_4/lrr3_datenblatt_de_2009.pdf.

2-29. Robert Bosch GmbH. Bosch Automotive Technology. Retrieved from http://www.bosch-automotivetechnology.us/en_us/us/component_us/SF_PC_DA_Predictive-Emergency-Braking-System_SF_CV_Driver-Assistance-Systems_511.html?compId=79.

2-30. Smartmicro. Automotive Radar. Retrieved from http://www.smartmicro.de/index.php/automotive-radar/overview-automotive-radar.

2-31. Delphi Automotive. Delphi Collision Mitigation System. Retrieved from http://delphi.com/shared/pdf/ppd/safesec/cms.pdf

2-32. Volvo Car Group (March 2013). Volvo Car Group Reveals World-First Cyclist Detection with Full Auto Brake in Geneva. Retrieved from https://www.media.volvocars.com/global/en-gb/media/pressreleases/48219.

2-33. Wüst, C. (February 2013). Auto Revolution: A Promising Future for Self-Driving Cars. Spiegel. Retrieved from http://www.spiegel.de/international/business/driverless-cars-are-a-reality-but-face-acceptance-obstacles-a-880716.html.

2-34. Ibeo Automotive Systems GmbH. Ibeo Technology. Retrieved from http://ibeo-as.com/fusion-system.html.

2-35. Car-to-Car Communications Consortium (October 2012). European Vehicle Manufacturers Working Hand in Hand on Deployment of Cooperative Intelligent Transportation Systems and Services [Press Release]. Retrieved from http://www.car-to-car.org/index.php?eID=tx_nawsecuredl&u=0&file=fileadmin/downloads/forum08/PressReleases/Press_release_on_MoU.pdf&t=1399062440&hash=f31221635e76425b171f04976134aa3f7d730c9c.

2-36. ITS Joint Program Office. Vehicle to Vehicle Communications for Safety. Retrieved from http://www.its.dot.gov/research/v2v.htm.

2-37. Muller, J. (March 2013). "No Hands, No Feet: My Unnerving Ride in Google's Driverless Car." Forbes. Retrieved from http://www.forbes.com/sites/joannmuller/2013/03/21/no-hands-no-feet-my-unnerving-ride-in-googles-driverless-car/.

2-38. Fleet Automation Forum (2013). Fleet Automation Forum: A Collaborative Model for Accelerating Commercialization and User Acceptance of Intelligent Vehicle Systems. Retrieved from http://sitemaker.umich.edu/cvpc/files/faf_whitepaper_csf.pdf.

2-39. Nowakowski, C., O'Connell, J., Shladover, S. E., and Cody, D. (2010). Cooperative Adaptive Cruise Control: Driver Acceptance of Following Gap Settings Less Than One Second. Proceedings of the Human Factors and Ergonomics Society Annual Meeting, 54(24), 2033-2037. doi: 10.1177/154193121005402403.

2-40. Van den Broek, T.H.A., Netten, B.D., Hoedemaeker, M., and Ploeg, J. (2010). The Experimental Setup of a Large Field Operational Test for Cooperative Driving Vehicles at the A270. Paper presented at the Intelligent Transportation Systems (ITSC), 2010 13th International IEEE Conference on September 19-22, 2010.

2-41. Helmer, T. (October, 2012). Highly Automated Driving–State of the Art and Future Challenges. Presented at International Task Force on Vehicle-Highway Automation.

2-42. Daimler AG (November 2012). New Assistance Systems: The Helpers in the Background. Retrieved from http://media.daimler.com/dcmedia/0-921-1549267-1-1549448-1-0-0-1549717-0-1-12759-614216-0-0-0-0-0-0-0.html?TS=1364849548507.

2-43. BMW Group (August 2011). Ready for Takeover! Taking BMW Automated Vehicle Technology to the Roadway. Retrieved from https://www.press.bmwgroup.com/pressclub/p/pcgl/pressDetail.html?title=ready-for-takeover&outputChannelId=6&id=T0119745EN&left_menu_item=node__2374.

2-44. Stoklosa, A. (2013). "We Go for a Ride in Audi's 'Piloted Driving' Autonomous A6 Avant Prototype." Car and Driver. Retrieved from http://blog.caranddriver.com/we-go-for-a-ride-in-audis-piloted-driving-autonomous-a6-avant-prototype-2013-ces/.

2-45. Honda Motor Europe Ltd. (April, 2008). LKAS/ACC Safety Features on the New Accord. Retrieved from http://www.honda-news.eu/cz/carnewsdetail/?id=899.

2-46. Nissan Motor Company. (January 2013). 2014 Infiniti Q50 Sports Sedan Unleashes Inspired Performance and Hospitality. Retrieved from http://infinitinews.com.

2-47. General Motors Corporation. (April 2012). Self-Driving Cars in Cadillac's Future. Retrieved from http://media.gm.com/media/us/en/cadillac/news.detail.html/content/Pages/news/us/en/2012/Apr/0420_cadillac.html.

2-48. Hoeger, R., Zeng, H., Hoess, A., Kranz, T., Boverie, S., Strauss, M., et al. (2011). HAVEit: Deliverable D61.1 Final Report. Regensburg, Germany Continental Automotive GmbH. Retrieved from http://www.explinovo.com/public/download/documents/HAVEit_212154_D61.1_Final_Report_Published.pdf.

2-49. Shladover, S. (2012). PATH Progress on Truck Platoons and Bus Steering Guidance. Presented at the TRB Road Vehicle Automation Workshop, Irvine, CA. Retrieved from http://onlinepubs.trb.org/onlinepubs/conferences/2012/Automation/presentations.pdf.

2-50. Tsugawa, S. (2012). Energy ITS: What We Learned and What We Should Learn. Presented at the TRB Road Vehicle Automation Workshop, Irvine, CA. Retrieved from http://onlinepubs.trb.org/onlinepubs/conferences/2012/Automation/presentations.pdf.

2-51. *Chan, E. (2012). An Overview of the SARTRE Platooning Project: Technology Leadership Brief, SAE* Paper 2012-01-9019. Presented at SAE Convergence, October 2012, Detroit.

2-52. Ministry of Land, Infrastructure, Transport and Tourism, Japan. (undated). Advanced Safety Vehicle: Realization of Secure and Safe Traffic Society by Harmonizing Humans and Vehicles. Retrieved from http://www.mlit.go.jp/jidosha/anzen/01asv/japanese/planning5.html.

2-53. Nothdurft, T., et al. (October 2011). Stadtpilot: First Fully Autonomous Test Drives in Urban Traffic. Fourteenth International IEEE Conference on Intelligent Transportation Systems.

2-54. Glaser, S., et al. (2012). Low Speed Automation, a French Initiative. Presented at Transport Research Arena, Athens, Greece. Retrieved from http://hal.inria.fr/docs/00/76/74/29/PDF/ArticleFawzi_TRA2012_paper_final_366_Word.pdf.

2-55. Alessandrini, A. (2012). CityMobil and other EC Projects. Presented at the TRB Road Vehicle Automation Workshop, Irvine, CA.

2-56. Mitchell, W., Borroni-Bird, C., and Burns, L. (2010). *Reinventing the Automobile: Personal Urban Mobility for the 21st Century.* MIT Press. ISBN: 9780262013826.

2-57. Williams, M. (October 2012). "Nissan Previews Self-Driving Car." *Computer World.* Retrieved from http://www.computerworld.com/s/article/9231978/Nissan_previews_self_driving_car.

2-58. Schoenherr, E. (2009). "Moving Future Convoy Operations with Convoy Active Safety Technologies (CAST)." *Accelerate Magazine,* July-September 2009. Retrieved from http://tardec.army.mil/Documents/AM0709_Moving_Future_Convoy_Ops.pdf.

2-59. Thiesen, B. (2011). Autonomous Mobility Appliqué System. Retrieved from http://www.sae.org/events/training/symposia/ivs/presentations/2011/bernard_theisen.pdf.

2-60. Lamm, R. (2012). Unmanned and Downrange. Technology Today. Available at http://www.swri.org/3pubs/ttoday/Summer12/pdfs/UnmannedandDownrange.pdf.

2-61. Main, D. (March 2013). Autonomous Cars Patrol Israeli Border. LiveScience. Retrieved from http://www.livescience.com/27949-self-driving-cars-israel.html.

2-62. Rio Tinto (July 2012). Rio Tinto's Driverless Trucks Achieve a New Milestone at Yandicoogina Mine under Mine of the Future™ Programme. Retrieved from http://www.riotinto.com.au/documents/120719_Rio_Tintos_driverless_trucks_achieve_a_new_milestone_at_Yandicoogina_mine_under_Mine_of_the_Future_programme.pdf.

2-63. Sakai, S. (March 2013). Introduction to Developments on Automation–MLIT. Presented at the iMobility Forum Automation Working Group meeting, March 7, 2013. Retrieved from http://icarsupport.eu/assets/Events/20130307-WS-on-Automation-Diamant-center-Brussels/1.a.3-MLIT-Japan.pdf.

2-64. Boyuan, C. (2013). Chinese-made Unmanned Vehicle Passes Freeway Test. China.org.cn. Retrieved from http://www.china.org.cn/china/2013-01/23/content_27773068.htm.

2-65. Jääskeläinen, J. (2013). Automated Driving from the European Perspective. Presented at the iMobility Forum Automation Working Group meeting, March 7, 2013. Retrieved from http://icarsupport.eu/assets/Events/20130307-WS-on-Automation-Diamant-center-Brussels/1.a.1-EC.pdf.

2-66. UR:BAN (2013). Urban Space: User Oriented Assistance Systems and Network management. Retrieved from http://urban-online.org/en/urban.html.

2-67. Johnson, T. (2012). Enhancing Safety Through Automation. Presented at the TRB Road Vehicle Automation Workshop, Irvine, CA. Retrieved from http://onlinepubs.trb.org/onlinepubs/conferences/2012/Automation/presentations.pdf.

2-68. Icarsupport (March 2013). Retrieved from http://icarsupport.eu/esafety-forum/.

2-69. Urmson, C. (August 2012). The Self-Driving Car Logs More Miles on New Wheels. Retrieved from http://googleblog.blogspot.com.au/2012/08/the-self-driving-car-logs-more-miles-on.html

2-70. Rupp, J., and King, A. (October 2010). Autonomous Driving—A Practical Roadmap. SAE Paper 2010-01-2335. SAE International, Warrendale, PA 2010.

2-71. Bartels, A. (March 2013). Automated Driving—State of the Art and Future Challenges. Presented at the iMobility Forum Automation Working Group meeting, March 7, 2013. Retrieved from http://icarsupport.eu/assets/Events/20130307-WS-on-Automation-Diamant-center-Brussels/3.a-Highway-Automation.pdf.

2-72. Toyota Motor Corporation (January 2013). Toyota Motor Corp. and Lexus Showcase Advanced Active Safety Research Vehicle at CES. Retrieved from http://pressroom.lexus.com/releases/toyota+lexus+advanced+active+safety+research+vehicle+ces+jan7.htm.

2-73. Hachman, M. (March 2012). "Ford Taking a Slow Road to Self Driving Cars." PC Magazine. Retrieved from http://www.pcmag.com/article2/0,2817,2401168,00.asp.

2-74. Bunkley, N. (July 2012). "The Car Without a Driver." Automotive News, July 16, 2012. Retrieved from http://www.autonews.com/apps/pbcs.dll/article?AID=/20120716/OEM06/307169920/1143#axzz2Os5XnrtM.

2-75. Smyth, G. (June 2012). Gary Smyth, Executive Director for North American Science Labs, General Motors, 2012 Driverless Car Summit Presentation and Interview. Available at www.auvsi.org.

2-76. Ford Motor Company. (June 2012). Ford Research Developing Intelligent System to Help Drivers Manage Stressful Situations on the Road. Retrieved from https://media.ford.com/content/fordmedia/fna/us/en/news/2012/06/27/ford-research-developing-intelligent-system-to-help-drivers-mana.html.

2-77. BMW Group. (September 2012). Safe and Efficient Driving in Tomorrow's Cities. Retrieved from https://www.press.bmwgroup.com/pressclub/p/pcgl/pressDetail.html?title=safe-and-efficient-driving-in-tomorrow%E2%80%99s-cities-in-the-ur-ban-research-initiative-specialists-from&outputChannelId=6&id=T01322 24EN&left_menu_item=node__2367n.

2-78. Toyota Motor Corporation. (September, 2012). Toyota Collaborative Safety Research Center Launches Seven New Projects and Four New Partnerships with Leading Academic and Research Institutions. Retrieved from http://pressroom.toyota.com/releases/toyota+csrc+seven+new+projects+four+partnerships+leading+academic+resea rch+institutions+sept2012.htm.

2-79. Smith, B. (2012). Automated Vehicles Are Probably Legal in the United States. Center for Internet and Society, Stanford University. Retrieved from http://cyberlaw.stanford.edu/publications/automated-vehicles-are-probably-legal-united-states.

2-80. Kalra, N., Anderson, J., and Wachs, M. (2009). Liability and Regulation of Autonomous Vehicle Technologies. (UCB-ITS-PRR-2009-28). Richmond, CA: University of California, Berkeley Retrieved from http://www.path.berkeley.edu/PATH/Publications/PDF/PRR/2009/PRR-2009-28.pdf.

2-81. Gasser, T. (2012). BASt Study: Definitions of Automation and Legal Issues in Germany. Presented at the TRB Road Vehicle Automation Workshop, Irvine, CA. Retrieved from http://onlinepubs.trb.org/onlinepubs/conferences/2012/Automation/presentations.pdf.

2-82. van Dijke, J., and van Schijndel, M. (2012). Citymobil, Advanced Transport for the Urban Environment. Transportation Research Record: Journal of the Transportation Research Board, 2324, 29-36. doi: 10.3141/2324-04.

2-83. Keane, A. (February 2013). Self-Driving Cars More Jetsons Than Reality for Google Designers. Bloomberg News. Retrieved from http://www.bloomberg.com/news/2013-02-06/self-driving-cars-more-jetsons-than-reality-for-google-designers.html.

2-84. Oonk, M., and Svensson, J. (March 2013). Automation in Road Transport Past Present and Future. Presented at the iMobility Forum Automation Working Group meeting, March 7, 2013. Retrieved from http://icarsupport.eu/assets/Events/20130307-WS-on-Automation-Diamant-center-Brussels/1.d-state-of-the-art.pdf.

Chapter 3
Cargo and Freight

Dan Williams, TRW Commercial Steering Systems

> The chief business of the American people is business.
>
> Calvin Coolidge

3.1 Introduction

The fundamental characteristic that differentiates commercial vehicles from passenger cars is their vocation. Commercial vehicles work for a living. Purchase and specification decisions are based solely on a business case. Professional drivers tend to be more capable and experienced. A general definition of commercial vehicles would include vocational trucks that operate on the highway and/or off-road. This chapter will deal with the applications of autonomous technologies to commercial vehicles operating on public roads and highways. As shown in Figure 3.1, self-guided means of transport have been profitably employed to move freight for some time.

Commercial vehicle duty cycles are as varied as the work they do. Fortunately, it is a good approximation that any commercial vehicle duty cycle is a bimodal combination of two extremes. First, and usually dominant, is a lane-keeping mode where the vehicle is maintaining lateral lane position at a high speed. Second is a maneuvering or docking mode, where the vehicle can experience large steering inputs but at relatively low ground speed. Since lateral acceleration is determined by the product of steering input and vehicle speed, both of these commercial vehicle modes are commonly characterized by low lateral acceleration. This allows these payload-carrying vehicles to have higher centers of gravity than the passenger cars they share the road with. Roughly two-thirds of all commercial vehicles are Class 8 semitractors that pull semitrailers with an almost exclusively lane-keeping duty cycle. The other third is comprised of all sorts of specialty vehicles, ranging from touring coaches that have roughly the same duty cycle as semitractors, to pick-up-and-delivery vans and transit buses that spend most of their time driving in urban environments, to school buses and refuse haulers whose urban duty cycles can be frequently punctuated by starts and stops, and yet these vehicles can be expected to see some time at high speed on highways.

Figure 3.1 Early autonomous freight haulers.

Any application of autonomous technologies to commercial vehicles is dependent upon the value created in these distinct duty cycles. Increased value is recognized through improved vehicle operating efficiency and decreased accidents. In fact, these two sources of value are coupled in that accidents are a cause of congestion, and congestion can cause accidents [3-1]. As shown in Figure 3.2, the total commercial vehicle miles traveled have more than doubled since 1980, while the total lane miles of roadway have increased by less than 5%. These trends obviously cannot be extrapolated indefinitely into the future. In the lane-keeping mode, autonomy can increase vehicle throughput on limited infrastructure, and correct implementation of autonomy can prevent accidents. Both of these sources of value are shared with passenger cars, although commercial vehicles might more directly value the positive effects, as large fleets are often self-insured.

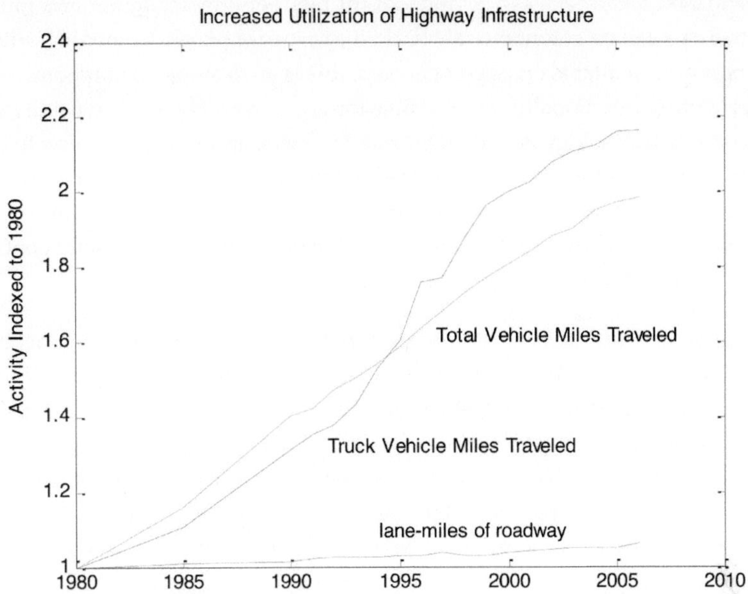

Figure 3.2 Increased utilization of highway infrastructure.

Commercial vehicles receive a less obvious but perhaps more important source of value from increased vehicle throughput. As throughput approaches the maximum on a given infrastructure, delays can be caused by events that would not be a factor for lower throughput. Therefore delays are more common as a given infrastructure is increasingly utilized. Delays inordinately cost commercial vehicles. Just-in-time (JIT) and lean manufacturing techniques have resulted in a virtual elimination of finished goods as well as raw material inventories in modern American factories. (The adoption of these manufacturing methods coincided with an increase in commercial truck fleet utilization, as seen in Figure 3.2.) In essence, plants' inventories are now housed in semitrailers. When the system works, it is a tightly integrated manufacturing system in which raw materials flow virtually continuously through a network of manufacturing operations at geographically distinct plants, resulting in finished goods flowing to the consumer. Traffic congestion causes uncertainty, uncertainty requires buffer inventory, and inventory is waste. Therefore commercial vehicles are particularly and specifically sensitive to reliability of transit times in ways that other vehicles are not. An issue with using autonomous vehicle technology to recover this value is the "free-rider" problem. Autonomous technology costs money, and as a commercial vehicle owner, you would prefer that everybody else invest to clear the roads for your benefit, and not have to do it yourself. In many ways this issue is particular to commercial vehicles that both disproportionately cause and are affected by congestion.

At the other end of the commercial vehicle's bimodal operating spectrum are low-speed maneuvers. Here value can be created in two ways. First, autonomous technology can provide more precise maneuvering than a typical operator. For example, an autonomous transit bus can successfully dock to a passenger loading platform so that passengers, including the disabled, can easily enter and exit the bus. Semitractor-trailer and multiple trailer combinations can be autonomously backed to a loading dock. Compared with the self-parking features now available on passenger cars, this is an example of autonomous technology providing functionality not possible for any driver. The value created by such autonomous vehicle technology is real, but limited to niche applications willing to pay for such precision. The second way value is created in low-speed maneuvers is by replacing the driver for repetitive tasks such as ferrying people to an amusement park entrance from a remote parking area. Truly driverless vehicles need to deal with other vehicles controlled by erratic drivers if they are going to share the road.

As autonomous vehicle solutions are considered in these applications, safety must be preserved. All autonomous vehicle manufacturers and their suppliers will risk product liability litigation in connection with vehicle accidents. Commercial vehicles carry cargo that is sometimes hazardous (or people) and have higher kinetic energy to dissipate at a given speed due to the added mass. Non-safety-related reliability is also critical for commercial autonomous vehicle applications. As we have seen, these vehicles are often on time-critical missions, and the failure of any component or system that prevents the timely completion of the mission will not be tolerated, even if it does not result in a safety issue. The business case and reliability are not completely decoupled, however. Many large commercial fleets are self-insured, so costs (or savings) from autonomous vehicle operation are directly realized. Furthermore, the most straightforward way to make safety critical systems more reliable is to add redundancy, which directly impacts implementation costs. We have seen that there are some similarities with other vehicles in how autonomous commercial vehicles might function, but the way this functionality is valued by commercial vehicle owners is different. Technological maturity has as much to do with autonomous commercial vehicle applications as functional value creation.

3.2 Technologies

3.2.1 Longitudinal Control Technologies

The technologies required to control autonomous vehicles can generally be divided into longitudinal and lateral functions. The powertrain, engine retarder, and brakes serve to accelerate or decelerate the vehicle. Commercial vehicle powertrains have become increasingly electronically controlled. Even "manual" transmissions are now automated, so that these systems can already support autonomous "gear-jamming" vehicle operation. Information regarding longitudinal control is available from the speedometer, and the simplest form of longitudinal control is cruise control that drives the engine based on a closed-loop error between the desired and measured vehicle ground speeds. The system components all can share information on the SAEJ1939 communication bus so that products from different suppliers can function as a vehicle system.

Brakes are a safety critical longitudinal technology. Pneumatic antilock braking systems (ABS) have been standard on commercial vehicles for some time. Electric braking systems (EBS) have been adopted in Europe to increase performance, adding a higher degree of control and blending to each individual wheel and eliminating delays associated with air actuation. Electric braking systems (brake-by-wire) and more conventional ABS also allow authority for a control intervention independent of the driver, making it a critical autonomous vehicle technology currently available. Current production EBS brakes are applied in parallel with several redundant pneumatic circuits for safety. They are an ideal actuator for autonomous vehicle braking and are currently accumulating a reliability field history.

For some time, radar-based sensors have been used to detect impending forward collisions and warn the driver. As shown in Figure 3.3, a variety of radar and other sensor technologies can combine to provide a more complete view around the vehicle. Using 2011 economics the Federal Motor Carrier Safety Administration calculated that such systems have between an 8- and 37-month payback period to fleets purchasing this technology [3-2]. Most fleets will invest with a one- to two-year payback, so this technology is currently viable in some applications. More recently, this information on impending collisions is used to make a decision that the vehicle should autonomously actuate its brakes if the vehicle is close to an unavoidable accident and there is no intervention from the driver [3-3]. This is perhaps the first application of an autonomous commercial vehicle function. Conventional ABS modulates the brakes to better achieve the drivers' indicated intention. In this collision avoidance application, the vehicle is performing a function not intended by the driver. This is a logical first step toward vehicle autonomy as the driver maintains safety critical lateral control, and the automatic intervention serves only to decrease the kinetic energy of the vehicle.

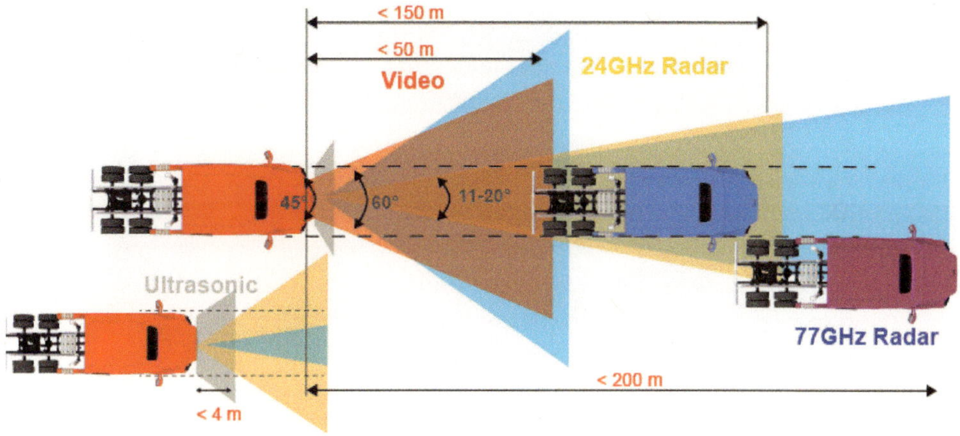

Figure 3.3 Combining sensors for more complete information.

More recently, information from radar and video-based technologies is used to estimate the distance to objects in front of the vehicle, allowing the cruise control to maintain a safe headway to a preceding vehicle, despite fluctuations in the preceding vehicle's speed. Similar to the collision avoidance system, the vehicle determines its own longitudinal control based on information from the environment independent of the driver. This is a step further toward autonomy as the vehicle now assumes control over accelerating inputs that can increase vehicle kinetic energy, rather than only dissipate it. This system has particular appeal to commercial vehicles, as the over-the-road duty cycle has a significant component of vehicle following. Karl Hedrick has shown how longitudinal dynamics can be controlled in a platoon formation using information passed from the lead vehicle to following vehicle [3-4].

3.2.2 Lateral Control Technologies

The video sensor introduced as part of the longitudinal control sensor suite has an even greater use in lateral control. Vehicle lane position can be determined from this video sensor, as well as future road curvature, as shown in Figure 3.4. SAFELANE demonstrated that a high-quality lane position can be determined with sensor fusion of the video sensor, radar sensor, and digital maps [3-5]. With such a sensor, the driver could be warned if a lane is breached. FMCSA has estimated that warning systems using such a video-based sensor to warn the driver of imminent lane departure have a 9- to 37-month payback, so these systems are likewise currently commercially viable in some applications [3-6].

Figure 3.4 Video processing of lane information.

Unlike EBS, which retains pneumatic redundancy in its production application, there is no good way to provide cost-effective redundancy and maintain the packaging advantages of steer-by-wire. Therefore, although steer-by-wire would offer the same implementation advantages to lateral control as EBS does for longitudinal control, it cannot be assumed to be available in the foreseeable future. However, from a control actuation standpoint, steer-by-wire systems are currently being used on the third axle of a three-axle vehicle. Such systems create value by reducing tire wear and improving vehicle maneuverability in niche applications [3-7]. Steering the rear axle of a tandem rear axle combination is a good initial application of this technology, as the three-axle vehicle is inherently relatively benign to failures. (Should the third axle be steered into an incorrect steering displacement, its subsequent yaw moment will be mitigated by the rigid unsteered second axle that immediately precedes the third axle in the rear tandem axle configuration.) As confidence in these electronically controlled systems increases with use over time, redundancies have been eliminated.

A second potential steering actuator is also accumulating a reliability history in the field. A computer-controlled electric motor has been integrated into the steering input driveline between the handwheel and the steering gear. The motor can be programmed to provide either an input to the steering gear or synthetic torque feedback (STF) to the driver (Figure 3.5). An early prototype form of this actuator was used in SAFELANE and was shown to provide value to the driver as a lane departure haptic warning device as well as a means for producing a correcting torque [3-8]. A well-conditioned position loop implemented on conventional electric power steering hardware has been shown to be an acceptable autonomous commercial vehicle steering actuator [3-9].

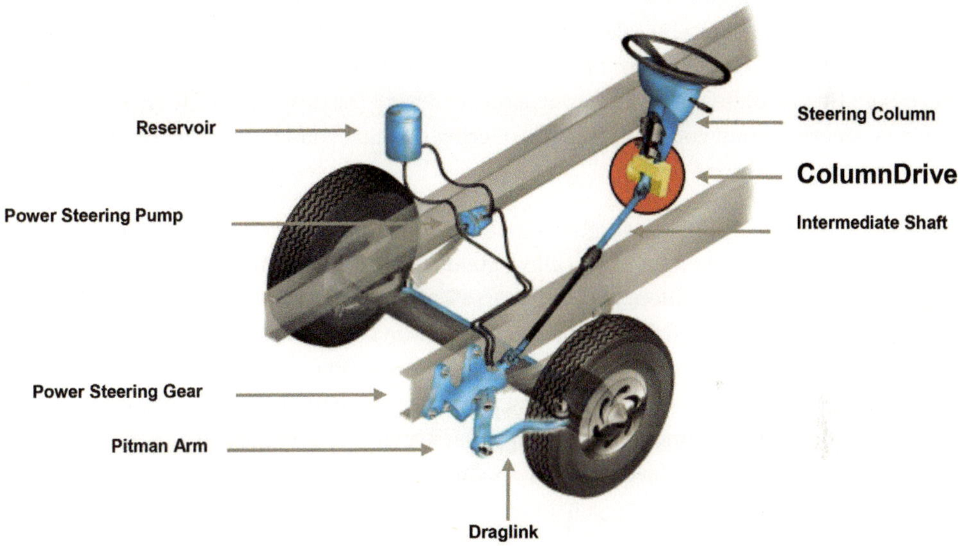

Figure 3.5 Synthetic torque feedback schematic.

In the current production form of STF, a control algorithm regulates the torque feedback to the driver [3-10]. The mechanical connection between the handwheel and roadwheel remains intact, so the redundancies to safely implement steer-by-wire are not required. The same hardware could be programmed to regulate the steering displacement input, becoming an autonomous vehicle control actuator. In fact, it is possible that such an actuator could be programmed to be in an autonomous (position control) mode, for example, by the driver manually throwing a switch, and revert to a driver control mode (torque control) when the driver grabs the steering wheel.

One of the features of a production STF is its ability to track the nominal steering position that results in a straight vehicle trajectory. The synthetic stiffness acts to return the handwheel to this identified position that is a slowly varying function of road crown, vehicle alignment, side wind, and so forth. Alternatively, this nominal position used in the torque control algorithm can be a function of the desired trajectory calculated from the lane position sensor. In this case, the synthetic stiffness pulls the handwheel toward the handwheel position that results in the desired trajectory. The amount of pull is dependent upon the stiffness parameter used and the amount of handwheel position offset (error). This desired dynamic handwheel position might or might not coincide with the nominal position that results in a straight running vehicle.

Knowing the current lane position and the desired lane position at some point in the future, it is a mathematical exercise to determine a desired trajectory connecting the two. Such a trajectory can be specified to be smooth, and to minimize lateral acceleration while converging in a specified time, or alternatively converging in a minimum time without exceeding a maximum lateral acceleration, or some more complicated trade-off between convergence time and lateral acceleration. Lane-keeping vehicle dynamics are very low frequency and small amplitude and therefore relatively linear. Inverse dynamics can be determined that result in a steering input associated with the desired vehicle trajectory. This calculated steering input could become the steering position command to the position control loop in the steering actuator.

Typically, the synthetic stiffness and the maximum allowable torque feedback are relatively low values in the production system to allow good feedback to the driver. With the dynamically changing nominal position, the synthetic stiffness and maximum allowable torque value can be increased, and the torque control algorithm begins to resemble a position control algorithm. As this occurs, the steering feel of the synthetic torque feedback is degraded. Therefore, the same lateral control hardware currently in commercial vehicle production can be programmed to achieve any particular point on a spectrum of performance from synthetic torque feedback to autonomous vehicle operation. The lateral control of vehicles is not strictly autonomous (position) control *or* manual (torque) control.

Global Positioning System (GPS) can be used to get a course indication of lane position. Current GPS functionality lacks fidelity required for precise lane positioning, but this key technology can play a part in a lane position estimated by sensor fusion. Furthermore, GPS can be used to overcome some of the shortcomings of the video-based sensor; for example, it could be combined with inertial sensors to provide a short-term estimate of vehicle position should the video sensor drop out due to snow or low visibility. Also, GPS could be used to map permanent conditions that the video sensor does not handle well, such as construction lanes, entrance and exits, and lane merges.

Another source of lateral lane position information is to place magnetic "nails" in the pavement and use an array of magnetic sensors on the vehicle to determine its location relative to the nails. This technology is available in all weather and visibility conditions (unlike the video-based sensor) and has very good resolution. Unfortunately, this technology requires infrastructure modifications that the others do not. Still, the magnetic sensor technology could have application in certain repetitive precision maneuvering applications where high-fidelity information is at a premium and the vehicle trajectory is very well defined.

The key longitudinal actuation technology (EBS) is currently in production, with pneumatic redundancies providing a margin for safety while field reliability is quantified and improved. Similarly, the key lateral actuation control technology (STF) is produced for a slightly different application. A parametric software change would be required for pure autonomous operation, but the hardware would be the same. Thus key autonomous commercial vehicle actuators are currently in the field, accumulating a reliability history, continually strengthening a business case dependent upon safety and reliability.

3.2.3 Advanced Communication Technologies

These technologies can be combined so that a commercial vehicle can perform its function without a driver. In some cases, autonomous vehicles can allow functionality beyond that achieved by a human driver that is limited in the ways information can be received and transmitted. Vehicle-to-vehicle (V2V) communication can be processed more efficiently by an autonomous vehicle. As commercial vehicles form a de facto platoon on an interstate highway, braking information from the preceding vehicle is transmitted to a human driver via brake lights. If the lead vehicle in a convoy brakes, the information is passed sequentially backward through the convoy as each driver observes the preceding vehicle's brake lights and engages their own brakes, thus signaling their immediate follower. An autonomous vehicle can share much higher bandwidth information with other vehicles and can be immediately alerted when the first vehicle in the convoy is engaging their brakes, independent of a specific vehicle's particular position.

The vehicle can also communicate with the infrastructure (V2I). In a strict sense, the lane position sensing technologies already discussed are an example of V2I, but in the future, V2I communication will be more sophisticated. For example, the presence of ice or standing water could be communicated to oncoming vehicles and speed automatically adjusted accordingly. Upcoming grade information can be transmitted to vehicles so that brake fade will not be experienced. Rather than (or in addition to) optical flares, a disabled vehicle will send out an electronic emergency signal, and the autonomous vehicle can change lanes well before; traffic back-ups are similarly anticipated.

Both V2V and V2I technologies are just as easily implemented in the higher volumes of passenger cars as commercial vehicles. Because of the potential for V2V and V2I to increase vehicle safety and operating efficiency, it is possible that they will create more value in the higher momentum commercial vehicle application, particularly when carrying hazardous payloads or passengers. These high-value payloads, however, could require a greater degree of communication integrity than the passenger car application. For example, 21st century pirates could use falsified V2V or V2I signals to stop high-value payloads where they could be readily commandeered, and terrorists could play havoc with hazardous payloads in a similar manner.

3.3 Vehicle Applications

3.3.1 Automated Docking

In many cases, potential autonomous commercial vehicle applications have predated emergence of the requisite technology in production. We are now seeing passenger cars parking autonomously. From a control standpoint, this is a very simple task compared with the commercial vehicle equivalent of backing-up an articulated vehicle to spot a trailer on a loading dock. In the late 1980s, Bernard Widrow, a Stanford University pioneer in nonlinear, adaptive and neural network control and others showed that a relatively simple neural net controller could successfully spot a standard configuration semitractor-trailer combination from any arbitrary starting condition [3-11]. Even an expert human driver might require several shifts from reverse to forward and back to reverse to properly position the trailer. Amazingly, the neural net controller could precisely spot the trailer using only reverse without shifting into a forward gear. In fact, this "truck-backer-upper" problem became a benchmark in the neural network arena, serving to establish the efficiency of various learning algorithms. By the mid-1990s, this technology was applied to double-trailer combinations [3-12].This application of neural network control to add an autonomous maneuvering mode to commercial vehicles allows operation not possible with human control and allows less-skilled drivers to perform the remaining tasks.

A promising initial application of autonomous vehicle technologies enables transit vehicles to perform precision maneuvers, such as docking. The University of California at Berkeley (PATH) did early research on the potential of this application, demonstrating that a prototype vehicle could precisely follow a trajectory defined by magnetic nails in the pavement. For the past ten years, Siemens has offered a system for sale that precisely guides a transit bus through a precision maneuver, such as docking, by using a video-based sensor and specialized roadway markings. In guidance mode, the vehicle is autonomously steered, and the driver maintains longitudinal control. "Optiguide" provides for autonomous vehicle control at speeds less than 70 kph and can track a trajectory to within 1 cm. See Figure 3.6. This allows buses to be safely driven very close to loading platforms. Fold-away loading ramps for disabled passengers are not needed, boarding time is reduced, and passengers experience lower lateral acceleration as approaches are smoother. The value proposition for Optiguide is not relative to conventional transit buses, but rather relative to rail-based rapid transit. With Optiguide, bus rapid transit claims the quality and efficiency of rail for less cost and using much existing infrastructure. Other examples of precision maneuvering include traversing a toll booth with minimal lateral clearance at moderate speed and automated movement through maintenance facilities.

Figure 3.6 Precision docking of transit vehicle.

3.3.2 Higher-Speed Headway Control and Platooning

Rapid transit has also explored the efficiencies of using dedicated lanes. It has been proposed that autonomous buses could precisely control their trajectory sufficiently to allow the buses to run on two narrow strips of concrete (rails) not much wider than the tires, so that a dedicated bus rapid transit lane co-exists with green space in an urban environment. Similar to precision docking, in this application a driver would remain present to control the longitudinal behavior of the vehicle. As before, the value of this system should be compared with rail rapid transit, and costs and infrastructure investment would be less.

Another autonomous vehicle application using a dedicated roadway has been suggested as a remedy to geographically constrained multilane roadways. Additional lanes cannot economically be added to some roadways due to a variety of reasons, such as proximity to the ocean, canyons, cliffs, and urban construction. In these cases it might be possible for a vehicle equipped with autonomous vehicle technologies to make use of the shoulder. Of course, the shoulder must retain its intended purpose as a repository for disabled vehicles, so V2V and V2I technology would be required to alert the driver of the autonomous vehicle that there is an impending disabled vehicle, and the autonomous vehicle needs to manually re-enter a conventional lane until the blockage is past. This application could have value for time-sensitive cargo (passenger, parcel delivery, JIT cargo) on frequently congested roadways. As the value proposition for these applications allows high-value freight carriers to invest in the technology to allow operation on reduced-width lanes, these vehicles not only experience more-efficient operation, but they free up space on conventional roadways for other vehicles. The bigger the cargo efficiency difference between the conventional lanes and reduced-width lanes, the more incentive there will be to invest in the instrumentation and thereby reduce conventional loading, so an economic equilibrium can be reached balancing installation costs with improved efficiency.

Finally, at high speeds there is potential platooning of commercial vehicles. Something like two-thirds of commercial vehicles participate in line-haul operation, or the movement of freight from point to point. These vehicles both cause and are affected by congestion, so using autonomy to improve the efficiency of this application is a potential benefit to the vehicle operator as well as other users of the interstate system. In principle, PATH demonstrated platooning of passenger cars using magnetic lane markers in 1997 [3-13], and snowplows in 1999. More recently, research emphasis has been on using video-based lane recognition to minimize infrastructure investment. The European project SARTRE, which includes a commercial vehicle manufacturer, has recently demonstrated platooning on unmodified motorways [3-14].

Using V2V communication convoys can be optimally organized so that interaction (i.e., entering and exiting the convoy) is minimized. With interactions minimized, and V2V communication allowing instantaneous transmission of vehicle states (such as braking), following distances could be dramatically decreased, easing roadway congestion and improving aerodynamic losses. Fuel is an over-the-road fleet's greatest operating expense. If a longitudinal control can significantly reduce following distances without dithering between acceleration and braking, fuel economy could be a more important influence on adoption of autonomous commercial vehicles than other vehicle types.

Taken a step further, these convoys of commercial vehicles could themselves run in dedicated lanes at higher speeds than currently allowed. The driver will likely not be replaceable in the foreseeable future in this application, as a driver will be required to control the vehicle before entering the platoon and upon exit. Therefore, the economic benefit of this application is strictly freight efficiency, and the higher speeds on specialized lanes would be a powerful advantage and avoid the "free rider" problem.

3.3.3 An Incrementally Adopted Integrated Future

At some point infrastructure changes will be required for many of the vehicle applications described. This investment decision is made easier if the vehicle technologies have been partially adopted toward ends justified by other sources of value. For example, it would be very hard for society to justify building a dedicated lane for high-speed commercial platooning, and once built, vehicles will have to install the required technology to take advantage of it. However, it can be seen that more vehicles will adopt intelligent cruise control as an assistance to the driver to unload the task of following a vehicle in moderate traffic requiring changes in speed from time to time. As confidence in the intelligent cruise control increases and perhaps with the addition of V2V technology, following distances might shrink, perhaps to avoid the temptation of passenger cars to encroach into the buffer between commercial vehicles with intelligent cruise control or to reduce aerodynamic losses. As following distances shrink, the lateral position control task is likely to become both mundane and more difficult, as the human driver will not be able to use the normal preview distance to determine lane position. As following distance decreases, more of the human driver's field of vision would comprise the preceding vehicle. At some point, steering is better achieved by a computer, perhaps using the forward-looking video stream from the lead vehicle in the platoon, possibly GPS positions of each following vehicle, and video cameras looking laterally to determine local lane position.

Once a critical mass of vehicles equipped with this technology are operating on the existing infrastructure, it will make sense on some routes to build a dedicated, limited-access express lane. Since all vehicles would be traveling on this lane at a high rate of speed, there is no need for passing slower vehicles going in the same direction. Traffic interaction would be limited, so that entering and leaving the platoon would only be possible at major cities or intersections of similar infrastructure, where long lengths of merge lanes would allow controlled entrance to exit from platoons with minimal acceleration/deceleration of the main body.

In many ways this vision for freight hauling autonomy resembles bus rapid transit, in how value is achieved by combining the point-to-point efficiency of rail with the flexibility and lower infrastructure required by wheeled vehicles. Extrapolating current infrastructure utilization trends into the future, a technological breakthrough will be required to maintain our industrial distribution capability while simultaneously supporting sustained economic growth on constrained infrastructure.

3.4 Summary

Commercial vehicles and their duty cycles have unique properties that affect adoption of autonomous technologies. Relative to passenger cars, commercial vehicles adopt technology more out of business case considerations. Freight hauling has always emphasized efficiency, but as JIT manufacturing has emerged, timeliness has become important. Freight haulers will readily invest in new technologies that improve efficiency and timeliness. Part of a compelling business case must include safety and reliability, and it is here that new technologies sometimes struggle to clear the investment hurdle.

Relative to passenger cars, commercial vehicles lack the scale required to support across-the-board adoption of new technologies. This situation is mitigated, however, by the fact that certain niche segments of the diverse commercial vehicle market offer strong motivation for local adoption. The adoption of autonomous technologies is aided as key technologies are currently in production supporting non-autonomous vehicle operation. With relatively minor changes in software or control, the same hardware could be used in autonomous vehicle operation. Because safety and reliability are such important components of the autonomous commercial vehicle value proposition, an incremental approach to the adoption of these technologies is likely.

As this hardware accumulates a field history, its functionality will start to approach autonomy. For example, when EBS brakes are actuated independent of the driver to avoid a collision with a stationary vehicle on the roadway, the vehicle can be considered autonomous in special circumstances. Similarly, when STF is applied in conjunction with a video-based lane position sensor to locate a dynamic steering center, the difference between driver assistance and autonomous driving is a matter of degree rather than kind. It is possible that as hardware associated with autonomous commercial vehicles is incrementally adopted, its functionality will move from predominately driver assistance to increasing autonomy.

We have already seen examples of how radar-based collision warning systems and EBS can combine to form a collision avoidance system, and STF and video-based lane-keeping can combine to form a vehicle that can autonomously control high-speed lateral lane position or low-speed trajectory. As these technologies combine with each other, and emerging V2V and V2I communication, they synergistically increase the value they provide and might eventually justify infrastructure modifications and thus widespread adoption.

3.5 References

3-1. Cambridge Systematics, "Traffic Congestion and Reliability," http://www.ops.fhwa.dot.gov/congestion_report/congestion_report_05.pdf.

3-2. Murray. D., Shackelford, S., and Houser, A. (February, 2009). "Analysis of Benefits and Costs of Forward Collision Warning Systems for the Trucking Industry," FMCSA-RRT-09-21, American Transportation Research Institute, Arlington.

3-3. Wabco press release, March 23, 2012, Louisville, KY. http://www.wabco-auto.com/nc/media/media_center/press_releases_wabco/press_releases_single_view/news-article/meritor-wabco-collision-mitigation-system-substantially-improves-commercial-vehicle-fleet-safety/.

3-4. Hedrick, J.K., McMahon, D., Narendran, V., and Swaroop, D. (1991). "Longitudinal Vehicle Controller Design for IVHS Systems," American Control Conference, Boston, June 26–28, 1991, pp. 3107-3112.

3-5. Beutner, A., et al. (2006). "D31.2.10 Final Report," PReVENT Consortium.

3-6. Houser, A., Murray, D., Shackelford, S., Kreeb, R., and Dunn, T. (February, 2009). "Analysis of Benefits and Costs of Lane Departure Warning Systems for the Trucking Industry," FMCSA-RRT-09-22, American Transportation Research Institute, Arlington, VA.

3-7. Williams, D. (2011). "Tire Wear Improvement by Steering a Third Axle," *SAE International Journal of Commercial Vehicles.* 4(1):1-12, 2011, DOI: 10.4271/2011-01-2148.

3-8. Beutner, A., et. al. (2006). "D31.2.10 Final Report," PReVENT Consortium.

3-9. Hingwe, P., Tai, M., and Tomizuka, M. (1999). "Modeling and Robust Control of Power Steering System of Heavy Vehicles for AHS." Proceedings of the 1999 International Conference on Control Applications, Hawaii.

3-10. Williams, D. (2009). "Synthetic Torque Feedback to Improve Heavy Vehicle Drivability." DOI: 10.1243/09544070JAUTO960. Proceedings of the Institution of Mechanical Engineers, Part D: *Journal of Automobile Engineering*, Vol. 223, No. 12, pp. **1517-1527.**

3-11. Nguyen, D., and B. Widrow, B. (1989). "The Truck Backer-Upper: An Example of Self-Learning in Neural Networks." DOI: 10.1109/IJCNN.1989.118723, International Joint Conference on Neural Nets, Washington, D.C.

3-12. Wang, B., Kinjo, H., Nakazono, K., and Yamamoto, T. (2003). "Design of Backward Movement Control for a Truck System with Two Trailers Using Neurocontrollers Evolved by Genetic Algorithms." *IEE Journal of Transportation. EIS,* Vol. 123, No. 5, pp. 983-990.

3-13. PATH, "Vehicle Platooning and Automated Highways," fact sheet. http://www.path.berkeley.edu/.

3-14. SARTRE. (2011). "First Demonstration of SARTRE Vehicle Platooning," press release. http://www.volvocars.com/intl/top/corporate/news/pages/default.aspx?itemid=239.

Chapter 4
Agricultural Applications

Marcel Bergerman, Carnegie Mellon University
John Reid, John Deere Co.
Eldert J. van Henten, Wageningen University
John Billingsley, University of Southern Queensland
Stewart Moorehead, John Deere Co.
Frits K. van Evert, Plant Research International
Sanjiv Singh, Carnegie Mellon University
Bradley Hamner, Near Earth Autonomy

> Without food, man can live at most but a few weeks; without it, all other components of social justice are meaningless.
>
> Norman Borlaug

4.1 Introduction

Agriculture, humankind's oldest and still its most important economic activity, is undergoing a revolution of equal or greater importance than the one it saw in the 18th, 19th, and 20th centuries. Then, steam and internal combustion engines provided field crop farmers with machines and vehicles for plowing, seeding, threshing, and harvesting that, for the first time in ten thousand years, allowed a small number of people to tend to thousands of acres of wheat, rice, soy, and other such crops. Today, the revolution is in the intelligence that we are embedding in those machines and vehicles, allowing them to drive autonomously down fields and orchards with a precision that is virtually impossible for a human driver to achieve. Such capability can be used to augment human workers and increase efficiency, for example, when apples are harvested from aboard a self-driving platform, or to replace workers in unhealthy or laborious activities, such as pesticide spraying or weed removal.

In this chapter, we discuss applications of autonomous vehicles to agriculture. We begin with a presentation of the social, technological, and commercial motivations for such applications and address the associated opportunities and challenges. We then proceed to describe four scenarios where autonomous vehicles have been successfully deployed in agricultural environments. We conclude with our thoughts on what's to come in the next decade and the years beyond that.

4.2 Agricultural Robotics

4.2.1 Challenges in Meeting Global Food Demand

There are significant challenges to close the gap in agricultural productivity to meet the demands for food, feed, fiber, fuel, and other uses to meet the world's need in 2050. While agriculture has lived up to the challenges in the past, these targeted increases in productivity will have to come in the face of more constraints facing agriculture—limited resources, lack of skilled labor, and land constraints, among other issues.

Total factor productivity (TFP) is the ratio of agricultural output to resource inputs employed in production. Significant achievements are going to be required from all of the factors that impact TFP if we are to achieve the production levels required in the future (Figure 4.1). Some predictions indicate a requirement for doubling of the output by 2050 [4-1], while managing sustainability will require growing the TFP of agricultural production systems from the current 1.4 to 1.75, consistently over the years.

Agricultural mechanization is one of the factors that contribute to increasing productivity. We have a global agricultural system that varies across crop types, production systems, and economic status. To make meaningful impact, autonomous vehicles must enable a significant increase in productivity through system automation, intelligent planning, or increased worksite productivity. There are two dimensions where this change is likely. One is where robotics makes a significant increase in machine productivity (e.g., automatic guidance has enabled more precision, faster speeds, reduced inputs). The other is where robotic systems result in breakthrough worksite solutions enabling system-of-systems levels of increased productivity inserted into the farm production system.

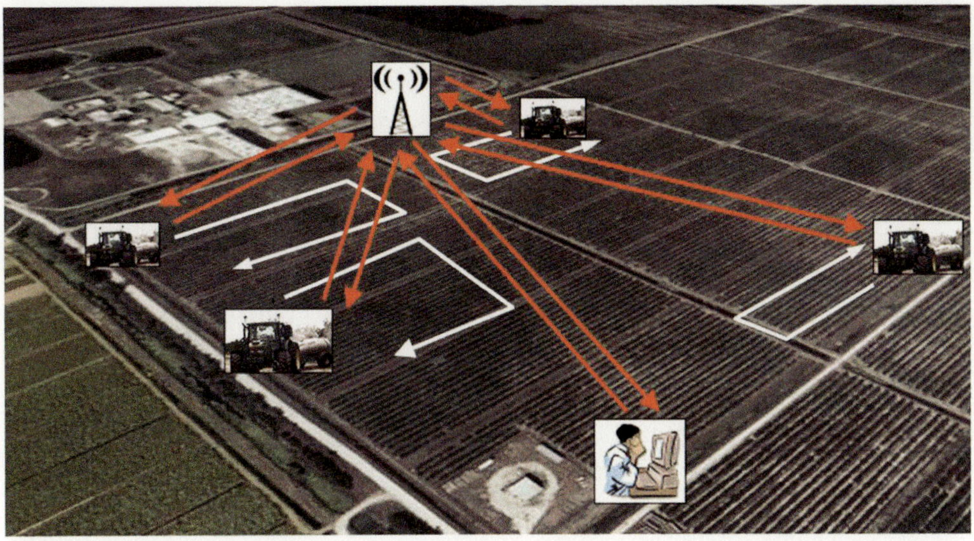

Figure 4.1 To fulfill the world's needs for food, feed, fiber, and fuel in 2050, the agricultural sector total factor productivity must increase 25% over current levels. Autonomous vehicles are one way to achieve this challenging goal. In this example, one person supervises four autonomous tractors.

4.2.2 From Mechanization to Autonomous Cyber-Physical Systems

Today's agricultural production systems are increasingly automated and becoming more permeated by information and communication technologies (ICT) that enable us to drive increased productivity through information collection, transfer, and management.

Precision agriculture, or precision farming, enables a systems approach for site-specific management of crop production systems. Its foundation is the concept of increased management of the inputs and documentation of production outputs leveraging geospatial data methods.

Precision farming results from one of the side effects of scaling agricultural mechanization through increasing the area that one person can manage, because machinery grew to a size that made it difficult to address opportunities for efficiency that might be driven by spatial, temporal, or system level effects [4-2]. A key technology enabler for precision farming resulted from the public availability of global navigation satellite systems (GNSS). GNSS started emerging in the mid-1990s and evolved to decimeter accuracy that could be used for mapping yield and moisture content or could provide course levels of control of machines. The opportunity space for robotics systems is greatly enabled by the evolution of GNSS systems that provide robust, high-accuracy localization.

Farmers have several types of machine systems available to them today that offer opportunities for varying degrees of machine control, including very precise spray nozzle or row-unit control. One of the more difficult elements not yet integrated on a large scale are precise maps or plans to guide spatially varying decisions, although a number of research concepts have been tested.

Site-specific maps can come from satellite imaging, aerial remote sensing, geographic information systems, field mapping, and other derivatives of these technologies that can provide information to drive a precision control system. Site-specific data sources like the aforementioned technologies provide a strong opportunity space for ICT and robotics solutions to increase productivity.

4.2.3 Challenges for Robotics and Automation to Drive Mechanization Toward Increased Productivity

The challenges that face robotics as an enabling technology fall primarily into social, technological, and business categories.

Social:

Robotics applications have to stand out as achieving key societal goals that enable human flourishing and should not be looked at as merely applications of novel technology. Agriculture has developed some very capable and productive systems that do very well and that set a high hurdle for new technologies to overcome. In addition to having value, robotics systems need to gain confidence as being robust, safe, and trustworthy production systems. Precision agriculture experiences in the past initially fell short because the setup or data management required significant nonproductive work that distracted the farmer from the task at hand.

Other major social issues relate to how robotics and automation applications fit within agricultural production systems. It is more acceptable to leverage robotics to close gaps in labor availability than to try to replace valued jobs. And the introduction of robotics technologies requires the management or rethinking of how work is accomplished in traditional systems in order to make it possible for the technology to work effectively.

The roles of human workers are usually not directly replaced completely by the current capabilities of robotic systems. Thus, the user experience of integrating robotics and automation may significantly change the skill-level requirements of workers to manage, support, or interact with the automated systems.

Technological:

Agriculture benefits from a number of enabling technologies that have made partial automation and machine system integration possible (GNSS and ICT capabilities were mentioned previously). But a major gap still exists in machine perception technologies that are required for environmental awareness of autonomous systems. A particular weakness is that the commercial maturity of these systems is low with significant capabilities still located in the research laboratory. Despite the major advances that have been made, the breadth of environmental conditions found in agricultural environments still presents a challenge to commercial deployment of sensing and perception systems. Other factors that may limit commercial execution include system design of products with rather large footprints and sophisticated agricultural machine forms, reliability, and cost.

Business:
The business basis for robotic systems in agriculture is one of the most critical challenges for the successful exploitation of the technology. Along with the strong customer expectation that the work gets done efficiently and effectively is the critical issue that the delivery of the system to the market is quite different from the purchase of standard agricultural equipment. It will take many years for robotics to achieve extraordinary levels of artificial intelligence capabilities, so the business must have a clear understanding of how the work gets done, how to monetize the work (perhaps differently than it is done today), and how delivery of the product to the customer is different (purchasing experience). The current "ag robotics" value chain is made up of mature companies embedded in the current agriculture paradigm that underestimate the complexity of how agricultural robotics is different from traditional robotics, mixed with robotics-focused players that are more interested in changing the world by selling robots than in deeply understanding how to serve the needs of the farmer. There are significant opportunities for entrepreneurial innovation in this space. Overall, the business model for robotics is unlikely to be the same as the model that has served agriculture so well over the recent past.

4.2.4 Opportunities Ahead for Robotic Systems

Despite these challenges, robotics technologies are really a strong enabler for increasing TFP in agriculture. And it is a natural evolution of our desire to leverage the capabilities of technology to advance human flourishing. As we march toward the population growth we expect to see by 2050, robotics-enabled agricultural systems will be key to providing solutions that contribute to feeding the planet.

4.3 Autonomous Vehicles in Agriculture

4.3.1 Arable Farming

Perhaps more than any other form of agriculture, arable farming has benefited from the mechanization that has occurred over the last 100 years. Large, efficient tractors and combine harvesters have allowed small numbers of farmers to cultivate thousands of hectares. Recently, advances in precision agriculture using GNSS have made the tasks of plowing, planting, and harvesting even easier and more efficient. The bounty afforded by this mechanization, along with great progress in plant genetics and fertilizers, has meant that arable farming has not been the subject of as much robotics research as have specialty crops.

While the creation of larger and more powerful machines has served arable farming well for the last century, a limit is being approached—the maximum size public roads and overpasses can support. To overcome this size barrier, many researchers have looked at creating systems of multiple tractors that drive in formation where only one vehicle is driven manually [4-3, 4-4]. These systems offer the advantage that a person is still present, and as such they have local contextual information of the local environment, compensating for any shortcomings in the autonomous perception system on the follower vehicles. In fact, the presence of the farmer in the cabin of the lead vehicle has allowed some to envision a future where the remote vehicles have no perception sensors for obstacle detection, assuming that the farmer is responsible for safeguarding the entire convoy. These coordinated vehicle systems provide a convenient solution to the need for a larger tractor, doubling or tripling the productivity of the individual farmer. They still suffer, however, from difficulty in transporting the system from field to field on public roads. These systems will most likely appear first on very large, contiguous farms where public road transportation is not an issue.

Over the course of a season, many passes are made over the fields to spray fertilizers, herbicides, and other chemicals that help improve the health of the planted crop and maximize yield. These practices have caused some concerns with chemicals entering water supplies or reaching other places where they may affect human health. Many techniques are thus being generated to spray smarter and minimize the amount of chemicals used on the fields. Systems based on GNSS are being combined with individually controlled spray nozzles to allow the spray to be turned on and off along portions of the boom as the vehicle crosses waterways and areas where spray is not desired. Some systems are combining computer vision to detect weeds and planted crop to spray herbicides only on weeds, rather than on the entire field, helping to reduce the amount of herbicides used and preventing damage to the planted crop.

Fully autonomous robot systems for arable farming have been created in research environments and have been demonstrated with great success [4-5]. However, a more common research topic has been to explore the concept of replacing a single large tractor with systems of small, autonomous vehicles [4-6]. The concept is that once labor is removed from the equation, it is much more beneficial for crop health to have small vehicles in the field. This offers the obvious benefit of lower soil compaction, but it also provides for greater system reliability, since the failure of a single vehicle does not stop all work as it would in a single large tractor solution. This solution, however, has several issues that will need to be overcome—logistical complexity, cost (it duplicates propulsion, structure, and perception sensors), and for tasks such as planting and harvesting, the challenge of carrying the large amounts of mass needed on a large farm.

4.3.2 Fruit Production

Fruit production is a notoriously labor-intensive activity. In the United States, for example, the cost of labor accounts for over 50% of the fixed cost to produce apples. Moreover, labor needs fluctuate significantly throughout the year—in the state of Washington, for example, the tree fruit industry employs 5,000 workers in January and 35,000 in October. To be profitable, farmers must continuously find skilled workers that accept seasonal jobs and who work long hours, in year-round weather, and at relatively low wages when compared to the industry or service sectors. There is a clear opportunity to introduce robotics and automation into tree fruit production to smooth out labor requirements, lower labor costs, and increase production efficiency. The opportunity is reinforced by the introduction, in the past two decades, of high-density planting architectures where fruit grows along "walls" of trees planted close to each other. Such environmental structure facilitates significantly robot navigation and localization, a fact that has not gone unnoticed by the robotics community.

To address this opportunity, we have introduced the autonomous prime mover concept (APM) [4-7]. The APM is a family of agricultural vehicles equipped with a common sensing and computing infrastructure that endows them with the basic capability to autonomously drive up an orchard row, turn around, and into the next. This seemingly simple behavior, however, is a powerful enabler for automating or augmenting most fruit production tasks, including mowing and spraying, pruning, thinning, tree training, and harvesting. APMs may range in size from small electric utility vehicles to large diesel platforms. We think of them as analogous to personal computers—devices of various sizes and power characteristics that run a common software base and can be connected to a variety of common external devices, as long as the appropriate hardware interfaces are present.

Over the past three years, we developed five APMs, four based on the Toro eWorkman MDE electric utility vehicle (Figure 4.2) and one based on the N. Blosi orchard platform. During this period, we identified, together with apple growers and extension educators, four APM operating modes and the functional and operational requirements that should drive the engineering phase of the APM family of agricultural vehicles. The four operating modes are the following:

Figure 4.2 The autonomous prime mover (APM) is a family of agricultural vehicles ranging from small utility vehicles, such as this one based on the Toro eWorkman MDE electric chassis, to large orchard platforms. Navigation in the orchard is accomplished with a horizontal laser mounted in front of the hood to find the rows of trees, steering and wheel encoders to measure distance traveled, and an onboard computer under the hood to run the control loop. Because of the structure of the environment in which they drive, APMs dispense high-accuracy GPS-assisted inertial navigation systems. This is fundamental to keep their cost low and make them accessible to growers.

1. **Transit mode:** Workers drive the vehicle manually (just as they would a regular car) from a storage location to the orchard at speeds up to 15 mph (24 km/h).
2. **Scaffold mode:** Workers stand on a platform onboard the vehicle and conduct tree operations such as pruning, thinning, tree maintenance, and harvesting while the vehicle drives autonomously down the row at speeds between 0.1 and 0.5 mph (0.2–0.8 km/h). The vehicle must be equipped with a user interface that allows workers to adjust speed and offset from the center and drive the vehicle manually via, say, a joystick (Figure 4.3).

Figure 4.3 (Top) Typical use of the APM in scaffold mode. Workers in a commercial orchard conduct green fruit thinning from onboard the vehicle, which moves slowly down the row. Tree maintenance trials conducted by the University of Pennsylvania indicate efficiency increases of up to 58% when workers use the APM compared with the same task done with ladders. (Bottom) Typical use of the APM in mule mode. Workers in an experimental orchard harvest the trees and place the fruit on bins carried by the vehicle. When the bins are full, the vehicle is commanded to the end of the row, where a crew manager carries the bins away with a tractor and replaces them with empty ones.

3. **Mule mode:** Workers walk along the row harvesting and placing the fruit in crates or bins on the vehicle or a trailer pulled by it. The vehicle autonomously follows the row at a safe distance behind or ahead of the workers; when it's full, it automatically goes to the end of the row where other workers can empty the crates/bins. In a more futuristic scenario, the vehicle would autonomously drive to the warehouse where the loaded bins are moved to storage and replaced by empty ones; the vehicle would then return to the orchard to join the harvest crew. In this mode, a simple two-button interface suffices for vehicle control: one button to start/stop motion and one to send the vehicle to the warehouse (Figure 4.3).
4. **Pace mode:** The vehicle drives autonomously an entire block without any worker onboard or nearby. Workers simply indicate to the vehicle the area to be covered, press "Start" on a suitable user interface, and wait until the vehicle completes the job. This mode is useful for operations such as block and selective spraying and cover crop mowing, and if the vehicle is equipped with the appropriate sensors, scouting the canopy for insects or disease and for crop yield estimation.

Based on these four modes, the functional and operational requirements identified are

- Smooth forward speeds from 0.1 to 5 mph for operation within crops as well as a higher speed mode of 10–15 mph while being driven to blocks
- Self-steered and easily operable by farm workers
- Fit into rows of modern high-density orchards without damaging fruit—ideally, at most 48 inches wide at 3 feet above the ground
- Stable on cross slopes of up to 10°
- Full workday endurance before refueling/recharging
- Operation in temperatures varying from 10 to 100° F
- Electric-battery powered for smooth/quiet operation as well as reduced emissions
- Equipped with a platform that can be widened as much as 8 feet

As of the publication of this title, APMs have logged a total of 330 km of autonomous driving in orchards and nurseries in the states of Washington, Pennsylvania, Maryland, and New York. We have focused so far on the basic autonomy (navigation and localization) functions, coping with a large variety of canopies and terrains, and the user interfaces needed for scaffold, mule, and pace modes. Future work includes increasing robustness to orchard operating conditions (e.g., better wheel slip control during turns, smoother row following at very low speeds), adding low-cost geo-location capability to the localization subsystem, and adding a safety layer that prevents the vehicle from running into people or bins in the row.

4.3.3 Protected Cultivation

Protected cultivation systems are used throughout the world as a powerful instrument to produce crops. They protect the crops from unfavorable outdoor climate conditions and pests and offer the opportunity to modify the indoor climate to create an environment that is optimal for crop growth and production, both in terms of quality and quantity. Protected cultivation is an intensive production method with high investment and operational costs, therefore permitting only the production of high-value fruit and vegetable crops like tomatoes, sweet pepper, and cucumber; flowers like roses, chrysanthemums, and gerberas; and many types of pot plants. In the past decades, in western societies, this type of production has been confronted with increasing size of production facilities, increasing labor costs, increasing problems with the availability of sufficiently skilled labor, health problems of the employees due to heavy and repetitive tasks, and growing competition on the national and international markets. Automation and robotics are considered to be a way to address these issues. Additionally, growing concerns with food safety call for such technology. And last but not least, a precision horticulture approach is increasingly adopted in which plants are treated on an individual basis, to improve quantity and quality of crop production while using resources as efficiently as possible. Given the current constraints on human labor, this has led to an even stronger call for automation and robotics.

A typical growing cycle of most crops in protected cultivation includes seeding, transplanting, planting, crop maintenance, harvesting, sorting, packing, cleaning, and preparation of the greenhouse for the next cycle. In all these steps, internal transport of men, machines, and produce plays a role. For single-harvest crops like lettuce this cycle is repeated every time after harvesting the crop. Multiple-harvest crops like tomatoes and roses last a year or even several years before a new crop is planted. Then crop maintenance, harvesting, sorting, and packing are repeated every few days during the whole production period of the crop.

A quick survey by one of the authors revealed that commercial solutions exist for the first and final phases of crop production [4-8]. They consist of automated systems largely based on mechanical engineering solutions, being not very flexible to variations in the working conditions, containing a small amount of sensors and not much "intelligence." The reason is simple. The tasks during these phases are relatively simple and focus on objects quite well defined in terms of location, size, shape, and color. In crop maintenance and harvest, the location, size, shape, and color are less well defined, and it is not surprising that for those steps of the production cycle hardly any commercial solutions for support or replacement of human labor exist. This kind of complexity calls for high-tech solutions using manipulation and gripping devices, a wide ranging of sensors, and advanced computing hardware and software that are able to deal with a varying and unstructured environment in a fast and effective way. These challenges have received considerable attention in research on agricultural robotics the past three decades.

Replacing human labor requires solutions for two main functions: first of all, a mobile platform for transportation of the system through the greenhouse; secondly, a device that performs the actual operation on the crop. Nowadays, many high-tech greenhouses are equipped with a rail type infrastructure that fulfills two purposes. Besides being a hot water heating pipe system, it acts as guidance and support for harvest and storage carts and other mobile equipment used during crop production. This rail largely simplifies the design and operation of mobile devices in greenhouses. The actual operation, for example harvest, requires steps such as the detection and localization of the fruit and assessment of its ripeness. For detection and localization, computer vision is commonly used. When fruit or flowers have distinct colors, visible light imaging is a feasible route. Yet in some cases, fruit, leaves, and stems have almost similar colors in the visible light spectrum, and different approaches need to be adopted. In the case of the cucumber harvester (Figure 4.4), the different reflection properties in the near infrared spectrum were exploited to detect green cucumbers in the green environment [4-9, 4-10]. Additional challenges for sensing are cluttering of fruit, occlusion of fruit by leaves and stems, and, last but not least, the disturbing effect solar radiation has on the illumination of the scene. Besides the vision system, this robot used a commercially available Mitsubishi RV-E2 manipulator and a gripper-cutter mechanism for removal of the fruit from the plants. Collision-free motion planning was used to steer the manipulator during harvesting [4-11]. The cutter consisted of a special thermal cutting device that was used to prevent the migration of viruses from one plant to the other during the harvesting process.

Figure 4.4 A cucumber harvesting robot.

Harvesting has received considerable attention in robotics research focused on protected cultivation. Yet, this is not the only time- and labor-consuming cultivation operation. In cucumber production, amongst others, removal of old, nonproductive leaves in the lower regions of the plant is a time-consuming task. Figure 4.5 shows the same platform used for harvesting, but in this case hardware and software are used for leaf removal from the plants [4-12].

Figure 4.5 A robot for automated leaf removal in a cucumber crop.

As of the writing of this chapter, no robotic system is commercially available for horticultural tasks in protected environments. Some challenges still remain. Current systems are too slow and not effective enough. Additionally, systems are usually being developed for a single crop. Yet, protected cultivation encompasses a multitude of different crops, and within crops different cultivars do exist that differ in terms of color, shape, and size, adding an additional degree of complexity.

4.3.4 Pasture Management

Weed control is an important element of pasture management on dairy farms. Weeds compete with grass for light, water, and nutrients, so that a pasture infested with weeds produces less feed and is of lower nutritional value than a weed-free pasture. In conventional farming, selective herbicides are used to control weeds without damaging the grass. In organic farming, no chemical herbicides are used, and weeds must be physically removed or destroyed, typically by hand. Here we report on the design, construction, and results of an autonomous vehicle to detect and control broad-leaved dock (*Rumex obtusifolius* L.) on grassland.

Broad-leaved dock is the most troublesome of grassland weeds. It is widely distributed in all temperate climates [4-13]. Infestation of weed-free pasture is virtually unavoidable and can take place, for example, when manure containing viable seeds is spread on the land, or through seeds in bird droppings. If the weed is not controlled, it will reach a high density and reduce grass yield by 10 to 40% (Figure 4.6) [4-14].

Figure 4.6 A dairy pasture near Wilnis, The Netherlands, infested with broad-leaved dock. The weed can be recognized by the characteristic brown stems, which bear small reddish-brown flowers.

Responding to a request by a study group of 17 organic dairy farmers in the heartland of Dutch dairy farming, Van Evert et al. [4-15] constructed a robot to detect and control broad-leaved dock in grassland (Figure 4.7). In discussions with the farmers, these design criteria were formulated:

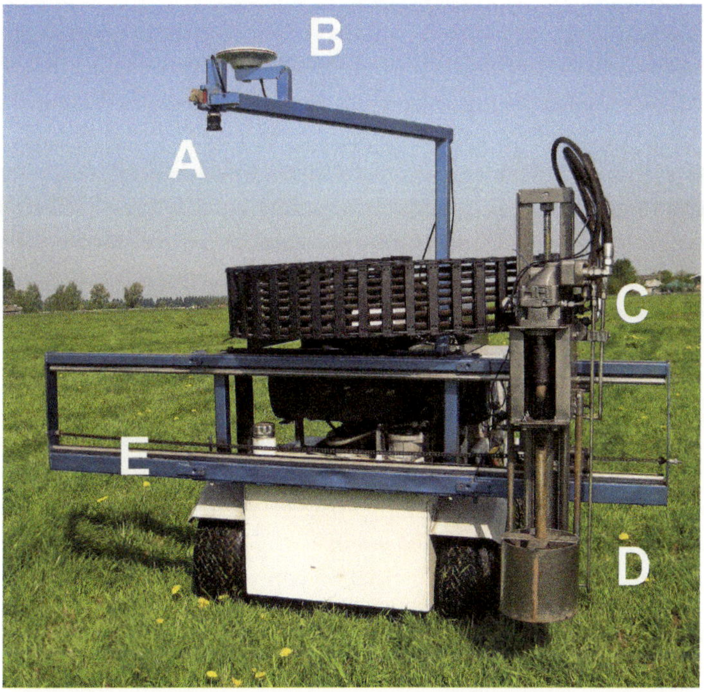

Figure 4.7 A robot to detect and control broad-leaved dock in grassland. Weeds are detected with a color camera (A). Navigation is GPS-based (B). A diesel engine (C) provides hydraulic power for the weeding implement (D) which can be moved laterally along the rail (E).

1. The robot will be deemed successful if it removes at least 70% of the weeds in a single pass.
2. The entire pasture must be searched.
3. The pasture is free of obstacles and the robot is allowed to travel everywhere, so high maneuverability is not required.
4. The robot has to be able to operate continuously for many hours.
5. The work rate is not critical.
6. Cost is an issue: Farmers indicated they would be willing to spend €2,500 ($3,250) p.a. on the robotic weeding system.
7. Safety need not be addressed in the prototype robot, because the robot will work in a fenced pasture where it should not encounter people, and because escape of the robot from the pasture is precluded by the drainage canals that separate the pasture from the road.

The above design criteria were met in the following way. Weeds and grass are both green; thus, a texture-based image analysis method was used to detect weeds [4-16, 4-17] (an early version of this method was demonstrated with a small-scale robot [4-18]). The weeder proposed by Riesenhuber was used [4-19]. The method consists of a rod weeder, which is driven vertically into the ground and which fragments the weed's taproot. Selection of the Riesenhuber method means that the robot needs to have a fairly large power source to drive the weeder; also, the mass of the robot needs to be sufficient to be able to push the weeder into the ground. This, combined with the weight of the power source, means the robot will have to be a sturdy vehicle. Traversal of a large, featureless pasture was achieved by using RTK-GPS to follow a predefined path consisting of parallel segments connected by half-circle turns at the ends. Skid steering was deemed sufficiently precise under the circumstances. The capability to operate continuously for many hours plus the energy needed to drive the weed chopper were met by installing a combustion engine. A 36-kW diesel engine that drives a hydraulic pump was selected because both are the current standard in agriculture. A custom-made platform was constructed to which four independently driven wheels were attached. The wheels are fitted with golf cart tires that provide traction on grass with minimal impact on the sod. Reduction gears ensure high torque, allow centimeter-precision forward movement, and limit maximum speed to a safe 3 mph.

The performance of the robot was quantified in an experiment on a commercial dairy farm. On two dates in October 2009, a pasture was selected that had been allowed to regrow after a grazing period. The grass was 10–20 cm tall. Broad-leaved dock density was high and variable. The diameter of individual weed plants varied between 0.1 and 0.5 m. On both days, the weather was highly variable, alternating between bright sunshine and overcast.

The accuracy of path following by the robot proved more than adequate. Deviations from the path were on the order of decimeters while following line segments but were much larger during turning. In all cases, the robot was able to quickly regain the path. Performance of weed detection and control was measured by logging all images taken by the system and using these to determine the number of weeds encountered, the number of false positives (weeding action in the absence of a weed), false negatives (no weeding action, even though a weed is present), and the positioning error of the weeder. In all, out of a total of 134 weeds, 93% were detected, while the false detection rate was about 6% (6 out of every 100 times that the robot indicated that it had detected a weed, there was in fact not a weed). Because earlier experiments had shown that 75% of weeds controlled did not grow back, the robot effectively removed 70% of weeds.

Project expenditure indicates that the sale price of a commercial version of this robot would be in the order of €50,000 (~$65,000), slightly below the number mentioned by Pedersen et al. [4-20], in part because the price of an RTK-GPS receiver has decreased. Depreciation, maintenance, and operating cost are estimated at €10,000 ($13,000) per year, or about four times the amount that the farmers' study group indicated they are willing to pay for robotic weeding. However, the capacity of the robot is such that one robot could be shared among five to ten farmers, either by direct collaboration between farmers or through a contract operator.

The robot meets design criteria 1–5 (70% removal, search entire pasture, tight maneuvering not required, many hours of continuous operation, work rate not critical). Criterion 7 (safety) has not yet been addressed. Criterion 6 (cost) has not been met for an individual farmer but could possibly be addressed by sharing a robot among several farmers. Also, a large part of the material cost is due to the RTK-GPS system and the power train (diesel engine, hydraulic pump, electronic valves) and might be lower in a commercial version. In short, it seems that a redesign will lead to a commercially viable robot.

Future work will seek to lower the cost of the system by (1) using electric motors (powered by a generator and a small combustion engine) and (2) using a low-cost GPS (and accepting a slightly larger positioning error).

As another example in this line of work, Figure 4.8 shows the Intelligent Autonomous Weeder [4-21, 4-22, 4-23]. It is a four-wheel steer, four-wheel drive platform to be used for autonomous weeding operations in arable farming. The platform combines dual GPS-based navigation with computer vision–based row following. The four-wheel steering construction offers supreme maneuverability, which is not only an advantage when precise operation within the crop is considered but also offers the opportunity of very compact head landturns.

Figure 4.8 The Intelligent Autonomous Weeder.

4.4 The Next Ten Years

The last ten years have seen the emergence of precision agriculture technologies, making use of GPS to precisely drive the tractor, spray chemicals, and map yield. These technologies have helped to improve yields and minimize inputs, providing great benefits to farmers. In contrast, the next ten years will be dominated by the increased intelligence and autonomy of agricultural vehicles. This trend will require the skill set of farmers to change, requiring greater comfort with technology from some and less skill from others as the machines become capable of doing more of the task autonomously. Greater technology on agricultural machines will also challenge manufacturers as they grapple with shorter technology life cycles and the desire of customers for interoperability of machines from different manufacturers.

The era of the smart machine has already started. The Guide Connect product from Fendt, where an unmanned tractor follows a conventional tractor in formation, won the Agrictechnica gold medal in 2011. Claas's AutoFill, which uses computer vision to control the spout in a combine or forage harvester into the grain cart, Trimble's WeedSeeker, which detects weeds and activates a spray nozzle to target the herbicide, and John Deere's Machine Sync, which drives the tractor while the combine is unloading into the grain cart, are all on the market today. These systems make use of several building blocks—computer vision, path planning, multivehicle coordination, and implement control—that are key to the creation of truly intelligent machines.

As the machines get smarter, they will be capable of collecting more and more information about the field, the crop, and their own internal "health." The next ten years will see an explosion in the amount of data collected and the ability to share this data between machines and farmers. New products such as John Deere's JD Link and FarmSight have laid the foundation for collecting information from machines, and the project iGreen [4-24] is exploring standard forms of information to allow sharing of data between machines from different manufacturers as well as exploring how farmers can share data with other farmers.

The rise of smart agricultural machines over the next decade will change the way food is grown. Along the way, many issues regarding safety, government regulations, interoperability, and information ownership will need to be solved. The benefits of smart agriculture, however, will provide the impetus to find solutions to these problems and help produce enough food to feed the ever-growing world population.

4.5 2020 and Beyond

Over the years, agricultural machines have become larger. The cause might be put down to the need to obtain the best productivity from their human drivers, a scarce farming commodity. At the same time, computing power and sensors have become cheaper, so that the necessary intelligence to add robotic features to the machines is much larger than their mechanical component. Perhaps we will very soon see a spate of small, autonomous agricultural robots, just as laptops have supplanted the mainframe.

Tasks such as selective spraying of weeds and of "chipping," removing weeds mechanically, do not require the robot machines to carry heavy loads, so a swarm of small machines can do the task as easily and with much greater redundancy than a single large device. Even in the case of harvesting, there is the present example of "helper" load carriers such as the cane-harvester's haul-out bin, so that collaborating smaller machines could perform the larger machine's task.

The infrastructure needed for central monitoring of a robotic team is rapidly being installed. Although there are many dead spots in the present mobile telephone network, governments have a commitment to provide broadband Internet services to blanket their countries, using satellites where necessary. An army of autonomous devices can be connected, long before 2020, with a farmer coordinating operations from the comfort of a farmhouse.

In the outback of Australia, cattle roam freely. They are regular visitors to watering places that are equipped with readers to monitor their RFID ear tags. As they gain access, they can be automatically weighed and the data sent back to a central repository. Automatic gates can segregate them when they reach a stage where they should be transferred to a feedlot for finishing and "harvesting." Although human intervention is needed for tagging the calves and performing the muster, the aim is once again to maximize the productivity of every farmhand by automation.

Great effort has been devoted to develop machine vision systems to aid in fruit picking. If this does not culminate in an efficient robotic fruit picker, it is possible to imagine retirees playing the role of "virtual backpackers," operating the mechanical end effectors as if playing a video game via TV screens in a retirement village.

Beyond harvesting, there are abundant sorting and grading operations that will benefit from computing and communication advances, with direct interaction with the data systems of the supermarket chains.

We envision a bright future for agricultural robotics and, in particular, the application of autonomous vehicles to agriculture. The dramatic advances in perception, computing, mobility, and manipulation achieved by the community in past years will enable manifold increases in a farm's total factor productivity. We are optimistic and will continue working toward helping to create such a future.

4.6 References

4-1. Global Harvest Initiative, 2012. http://www.globalharvestinitiative.org/. Accessed March 30, 2012.

4-2. Sevila, F., and Blackmore, S. (2001). "Role of ICTs for an Appropriate World Market Development." Presentation at the 12th Members Meeting, Club of Bologna, Bologna, Italy, November 18–19, 2001. http://www.clubofbologna.org/ew/documents/Proc2001.pdf.

4-3. Fendt GuideConnect: Two Tractors—One Driver, 2012. http://www.fendt.com/int/press_pressreleases_4569.asp. Accessed on March 30, 2012.

4-4. Reid, J.F. (2011). "Robotics and Automation as an Enabler to Agricultural Systems Productivity." Keynote Address, Robobusiness, Boston.

4-5. Pilarski, T., Happold, M., Henning, P., Ollis, M. (2002). "The Demeter System for Automated Harvesting." *Autonomous Robots*, 13(1): pp. 9-20.

4-6. Blackmore, B. S., Stout, W., Wang, M., and Runov, B. (2005). "Robotic Agriculture –The Future of Agricultural Mechanisation?" *5th European Conference on Precision Agriculture.* Stafford, J.V. (ed.), The Netherlands, Wageningen Academic Publishers, pp. 621-628.

4-7. Singh, S., Bergerman, M., Cannons, J., Grocholsky, B., Hamner, B., Holguin, G., Hull, L., Jones, V., Kantor, G., Koselka, H., Li, G., Owen, J., Park, J., Shi, W., and Teza, J. (2010). "Comprehensive Automation for Specialty Crops: Year 1 Results and Lessons Learned." *Journal of Intelligent Service Robotics, Special Issue on Agricultural Robotics*, 3(4): pp. 245-262.

4-8. Van Henten, E.J. (2006). "Greenhouse Mechanization: State of the Art and Future Perspective." *Acta Horticulturae*, 710: pp. 55-70.

4-9. Van Henten, E.J., Hemming, J., Van Tuijl, B.A.J., Kornet, J.G., Meuleman, J., Bontsema, J., van Os, E.A. (2002). "An Autonomous Robot for Harvesting Cucumbers in Greenhouses." *Autonomous Robots*, 13: pp. 241-258.

4-10. Van Henten, E.J., Van Tuijl, B.A.J., Hemming, J., Kornet, J.G., Bontsema, J., van Os, E.A. (2003). "Field Test of an Autonomous Cucumber Picking Robot." *Biosystems Engineering,* 86: pp. 305-313.

4-11. Van Henten, E.J., Hemming, J., Van Tuijl, B.A.J., Kornet, J.G., Bontsema, J. (2003). "Collision-free Motion Planning for a Cucumber Picking Robot." *Biosystems Engineering,* 86: pp. 135-144.

4-12. Van Henten, E.J., Van Tuijl, B.A.J., Hoogakker, G.-J., Van der Weerd, M.J., Hemming, J., Kornet, J.G., Bontsema, J. (2005). "An Autonomous Robot for De-leafing Cucumber Plants Grown in a High Wire Cultivation System." *Biosystems Engineering,* 94: pp. 317-323.

4-13. Cavers, P.B., and Harper J.L. (1964). "*Rumex obtusifolius* L. and *R. crispus* L." *Journal of Ecology,* 52: pp. 737-766.

4-14. Courtney, A.D. (1985). "The Role and Importance of Docks in Grassland." *Agriculture in Northern Ireland,* 59: pp. 388-392.

4-15. Van Evert, F.K., Samsom, J., Polder, G., Vijn, M., van Dooren, H.-J., Lamaker, A., van der Heijden, G.W.A.M., Kempenaar, C., van der Zalm, T., Lotz, B. (2011). "A Robot to Detect and Control Broad-leaved Dock (*Rumex obtusifolius* L.) in Grassland." *Journal of Field Robotics,* 28: pp. 264-277. DOI: 10.1002/rob.20377.

4-16. Polder, G., Van Evert, F.K., Lamaker, A., De Jong, A., Van der Heijden, G.W.A.M., Lotz, L.A.P., Van der Zalm, T., Kempenaar, C. (2007). "Weed Detection Using Textural Image Analysis." *6th Biennial Conference of the European Federation of IT in Agriculture* (EFITA), Glasgow. Available online at http://edepot.wur.nl/28203. Accessed on March 30, 2012.

4-17. Van Evert, F.K., Polder, G., Van der Heijden, G.W.A.M., Kempenaar, C., Lotz, L.A.P. (2009). "Real-time, Vision-based Detection of *Rumex obtusifolius* L. in Grassland." *Weed Research,* 49: pp. 164-174. DOI: 10.1111/j.1365-3180.2008.00682.x.

4-18. Van Evert, F.K., Lamaker, A., De Jong, A., Groendijk, E.J.K., Van der Heijden, G.W.A.M., Lotz, L.A.P., Peterson, S.L., Polder, G., Van der Zalm, T. (2009). "Sietse III: Navigation and weed detection using textural image analysis." In Oksanen, T., & van Henten, E. (eds.), *Proceedings of the Fourth Field Robot Event in Hohenheim, Germany.* Farm Technology Group, Wageningen UR, Wageningen. ISBN 978-90-8585-480-7. Available online at http://www.fieldrobot.nl/downloads/Proceedings_FRE2006.pdf. Accessed on March 30, 2012.

4-19. Böhm, H., and Finze, J. (2004). Überprüfung der Effektivität der maschinellen Ampferregulierung im Grünland mittels WUZI unter differenzierten Standortbedingungen. [Testing the effectiveness of mechanical control of docks in grassland with the WUZI under a variety of conditions.] Available online at http://orgprints.org/4165/01/B%C3%B6hm-B%C3%96L-Pflschutz-2004.pdf. Accessed on March 30, 2012.

4-20. Pedersen, S.M., Fountas, S., Have, H., and Blackmore, B.S. (2006). "Agricultural Robots—System Analysis and Economic Feasibility." *Precision Agriculture,* 7(4): pp. 295-308.

4-21. Bakker, T., Wouters, H., Van Asselt, K., Bontsema, J., Tang, L., Müller, J., Van Straten, G. (2008). "A Vision Based Row Detection System for Sugar Beet." *Computers and Electronics in Agriculture*, 60 (1): pp. 87-95.

4-22. Bakker, T., Van Asselt, K., Bontsema, J., Müller, J., Van Straten, G. (2010). "A Path Following Algorithm for Mobile Robots." *Autonomous Robots*, 29 (1): pp. 85-97.

4-23. Bakker, T.,van Asselt, K., Bontsema, J., Müller, J., van Straten, G. (2011). "Autonomous Navigation Using a Robot Platform in a Sugar Beet Field." *Biosystems Engineering*, 109 (4): pp. 357-368.

4-24. iGreen Project, 2012. http://www.igreen-projekt.de/iGreen/index.php?id=47&L=3. Accessed on March 30, 2012.

Chapter 5
Beyond Rosey: Consumer Robots in the 21st Century

Brian Yamauchi, iRobot

> Any sufficiently advanced technology is indistinguishable from magic.
>
> Arthur C. Clarke

5.1 Chapter Summary

In 1962, the animated series *The Jetsons* introduced the world to Rosey the Robot, a quasi-humanoid robot maid who took care of all of the Jetsons' cooking, cleaning, and housekeeping needs. In 2002, iRobot introduced the Roomba® vacuum cleaning robot (Figure 5.1). As of 2014, over 10 million Roomba robots have been sold. Today, the world is exactly halfway between *The Jetsons'* television debut and its 2062 setting, and robots can be found in homes around the globe. However, those robots bear little resemblance to Rosey. Today's consumer robots are special-purpose devices meant to perform a particular task, such as vacuuming floors, cleaning swimming pools, mowing lawns, or clearing gutters.

Figure 5.1 iRobot Roomba® 780 robot vacuums the floor in a home environment. (courtesy of iRobot)

What will tomorrow's consumer robots look like? Will they be smarter versions of the robots that exist today? Or will we see humanoids like Rosey in everyday households? This chapter surveys the consumer robots available today and looks forward to the robotic technologies and applications that are likely to change the world over the next century.

5.2 Introduction

Science fiction is becoming reality. From smartphones to cloning to space tourism, the dreams of yesterday are the technologies of today. However, the ways in which those dreams are realized often surprise even their own inventors. The researchers who invented the ARPANET, later to become the Internet, most likely would not have imagined that one of its major uses would be to allow millions of teenagers to track each other's relationship status on a minute-by-minute basis. Likewise, many futurists expected lunar bases and trips to Mars by the turn of the 21st century, but few expected space tourism to advance faster than NASA's manned space program.

Bearing this in mind, this chapter contains speculations on how consumer robotics may evolve in the coming decades. For most of the relatively brief history of the robotics field, the idea of a "consumer robot" was an oxymoron. Robots were expensive industrial devices used on assembly lines or, more recently, expensive bomb disposal robots used by the military and law enforcement agencies. But just as computers went from being multi-million-dollar behemoths used by the government and large corporations to devices that fit in every child's pocket, robotics technology has finally come down in price to the point where mass market consumer robots are possible.

Robots can be useful to the consumer in a number of different ways. First, an autonomous robot can perform unpleasant tasks, such as vacuuming or mopping. Second, a teleoperated robot can allow a user to be virtually present at a remote location, not merely observing the events but physically interacting with the world. Third, semiautonomous robots can assist the elderly and handicapped in performing tasks that they are no longer able to do on their own. These three general classes of robots can be applied to a wide range of tasks and environments, which we will examine in this chapter.

5.3 Indoor Home Robots

Over a decade ago, I joined iRobot and was working on the autonomous navigation system for an industrial floor-cleaning robot. Some members of the team advocated for the idea of "small, round robots" using insectlike behaviors to reactively navigate around home environments. The idea of using insectlike behaviors for robot control was championed by Rodney Brooks [5-1], director of the MIT Computer Science and Artificial Intelligence Lab (CSAIL). Earlier robots had tended to rely on rigid, top-down, symbolic planning systems for autonomy, and most had very limited capability to operate robustly in complex, unstructured environments. In contrast, behavior-based robotics provides a bottom-up methodology for building autonomous robots. Research by Brooks and others [5-2, 5-3] showed how biologically inspired approaches could generate adaptive behavior with animal-like robustness in real-world environments.

Together with Colin Angle (iRobot's CEO) and Helen Greiner, Brooks founded iRobot in 1990 to apply this approach to commercial robot products. iRobot used behavior-based robotics to develop the iAdapt control software for Roomba products. This software enables Roomba robots to clean effectively in a wide range of homes without any prior knowledge about the arrangement of those environments. Simple behaviors such as spiraling outward and inward and following walls and the contours of obstacles provide a robust capability for navigation in unknown environments.

In 2002, the Roomba robot was launched to the public, and it became an instant hit. Millions of the robot vacuums were sold around the world in the Americas, Europe, and Asia. Overnight, the Roomba robot became the world's most successful autonomous robot, by several orders of magnitude. In 2005, iRobot followed the Roomba with the Scooba® floor-washing robot (Figure 5.2). Later, other companies, such as Samsung and Neato, began producing their own robot vacuum cleaners, with a variety of different sensors and autonomy algorithms.

Figure 5.2 iRobot Scooba® 230 floor-washing robot cleans bathroom floor. (courtesy of iRobot)

In the future, most housecleaning tasks will be performed by robots. Though reactive behaviors work well to cover floors, future household robots will need more sophisticated navigation capabilities to drive efficiently from one place to another. Advances in simultaneous localization and mapping (SLAM) technology allow a robot to build a map of the environment at the same time that it uses that map to determine its own location [5-4]. Using this technology, users will be able to specify destinations using a touchscreen map or simply say, "Go to the kitchen," and the robot will drive to the appropriate location.

Farther in the future, robots will need additional intelligence and more-sophisticated manipulation capabilities to perform tasks such as collecting and storing objects. For example, a robot could learn to recognize different objects and place them in the appropriate places (e.g., books on bookshelves, clothes in the hamper, trash in the wastebasket). Recent advances in computer vision include the ability to learn to recognize objects in a wide variety of real-world environments [5-5]. Research projects such as ARM-S (Advanced Robotic Manipulation–Software), funded by the Defense Advanced Research Projects Agency (DARPA), have developed robotic capabilities using dexterous manipulators to grasp and manipulate a wide range of objects–from light bulbs to power drills to door handles. Willow Garage has demonstrated their PR2 (Personal Robot 2) performing complex manipulation tasks such as folding laundry. Right now, the robotic hardware needed to perform such tasks costs hundreds of thousands of dollars. However, costs will fall in the future, and companies are working to develop lower-cost robots that may be more suitable for operation in home environments, with arms that are flexible and compliant rather than stiff and rigid.

5.4 Outdoor Home and Yard Robots

Robot lawnmowers, such as the Husqvarna Automower, the Friendly Robotics RoboMow, and the John Deere Tango ES5, have been available for the last decade. All of these robots require the owner to bury a wire (or attach it with spikes) around the perimeter of the lawn, as well as around any areas that should be avoided (e.g., flower gardens). These robots drive in random directions and change directions when they detect the current in the perimeter wire in order to remain within the contained area. All of these robots are battery powered and automatically dock with a charging station when their batteries run low. Using this random coverage strategy, these robots eventually mow the entire lawn. Robot lawnmowers have achieved moderate levels of commercial success, selling tens of thousands of units. However, high prices and the difficulty of installing the perimeter wire have limited adoption as compared with robot vacuum cleaners.

Pool-cleaning robots like the iRobot Verro® (Figure 5.3), the Aquabot™, and the Polaris robotic pool cleaner have become commonplace in homes with swimming pools. These robots drive around the bottom of the pool and climb up the walls, scrubbing the surface and gathering leaves and other debris. The Solar-Breeze solar-powered pool skimmer robot floats on the surface of the water and continuously drives around collecting floating debris. In the future, outdoor robots may also help homeowners with window washing and garden tending.

Figure 5.3 iRobot Verro® 500 pool-cleaning robot cleans swimming pool wall. (courtesy of iRobot)

5.5 Telepresence Robots

Videoconferencing has become commonplace, first in the workplace and then through consumer applications such as Skype. These apps allow people to see and talk to people at distant locations, but they suffer from some limitations. First, the user is unable to directly control the position of the remote camera. In business videoconferencing systems, cameras are typically locked down to a fixed position in a single room. For smartphones, remote users have no ability to change the camera position, other than by asking the person holding the phone to move it for them. Second, the remote user has no physical presence in the space for others to interact with. People who are participating remotely have less of a presence within the meeting and are often ignored or forgotten. Finally, remote users have no way to physically interact with the environment through a videoconference call, for example, by manipulating objects or using tools.

Telepresence robots allow users to experience remote locations in a richer and more visceral manner than videoconferencing. Using a telepresence robot, the remote user can move through the environment, driving to different rooms and locations, and seeing events from multiple viewpoints. A robot attending a meeting is more of a full participant than someone attending via videoconferencing. A telepresence robot can inspect a factory floor and engage employees in conversations. Telepresence robots allow remote users to perform gestures to point at objects in the environment and emphasize verbal points. Future telepresence robots will be equipped with manipulator arms that allow remote users to pick up and examine objects, to demonstrate physical actions, to use tools, and to open doors.

Telepresence robots may be either directly teleoperated or semiautonomous. Under direct teleoperation, a remote user drives the robot by sending low-level commands to move forward and backward and to turn left and right–much like controlling a character in a first-person shooter video game. Thousands of iRobot PackBot®, iRobot SUGV (Small Unmanned Ground Vehicle), and Foster-Miller Talon robots have been used in this manner on the battle-fields of Afghanistan and Iraq. These EOD (explosive ordnance disposal) robots are used to inspect and disarm suspected roadside bombs and IEDs (improvised explosive devices) from a safe distance. They have saved hundreds of lives and have revolutionized the way in which the military responds to the asymmetric threats of unconventional warfare.

These military robots are much too expensive for consumer use, often bearing six-digit price tags, but commercialized versions of telepresence have become available at much lower price points. The AnyBots QB and vGo telepresence robots are examples of low-cost, directly tele-operated robots. These robots have enabled people to remotely attend business meeting, school classes, and even weddings.

Although directly teleoperated robots can be inexpensive and useful, they suffer a number of key limitations. First, substantial bandwidth is required to continuously stream high-quality video, though this is becoming more available as high-speed Internet connections (both wired and wireless) become more commonplace for streaming high-definition movies and television. Second, any latency between the time a user sends a command, the time the robot executes the command, and the time the user sees the effects of the command can make controlling the robot difficult. This is a hard problem, but one that is being addressed to a certain extent in multiplayer first-person shooter games like Modern Warfare and Battlefield 3 and cloud gaming services like OnLive. In gaming, as in telepresence, any significant lag puts the user at a severe disadvantage.

The third and most challenging problem is the lack of situational awareness. The human eye has a horizontal field-of-view of 175 degrees and a vertical field-of-view of 160 degrees. This is far greater than any conventional camera lens. Even wide-angle lenses typically have a field-of-view of less than 90 degrees. Fish-eye lenses can have 180-degree fields-of-view but produce images with extreme distortion. Software can remove this distortion by remapping the image to a rectilinear perspective but at the cost of reduced resolution. Moreover, people can move their eyes and turn their heads quickly and easily to take in their surroundings. In contrast, the pan/tilt mechanisms on most remote cameras tend to be slow, and the act of controlling the camera angle using a joystick can be difficult and nonintuitive, especially in the presence of latency.

One approach to making camera control more intuitive is to combine head-mounted displays with a head-aimed camera gimbal. iRobot worked with Chatten Associates to integrate Chatten's HARV (Head-Aimed Remote Viewer) with iRobot's PackBot and Warrior® robots (Figure 5.4) [5-6]. The HARV includes a high-speed, high-precision, three-axis camera gimbal that was mounted on the robot. To use the HARV, the operator wears a headset that includes a head-mounted display and a head tracker incorporating a three-axis MEMS (microelectro-mechanical system) gyro. To look around, the operator simply turns his or her head, and the gimbal aims the camera in the corresponding direction. This provides far greater situational awareness with far more intuitive control and provides a highly immersive telepresence experience. However, this approach requires even more bandwidth and even lower latency to work effectively.

Figure 5.4 iRobot Warrior® prototype equipped with Chatten Associates Head-Aimed Remote Viewer (HARV) for immersive telepresence. (courtesy of iRobot)

When robots are teleoperated locally over local, high-bandwidth, low-latency communications channels, such as the wireless networks used for EOD robots, these problems can be minimized, but consumer telepresence robots are typically operated over the Internet with limited bandwidth and variable latency. As a result, these robots frequently collide with people and objects and require assistance from bystanders or full-time, on-site human handlers [5-7]. While this may be acceptable in a school, for example, it would be unacceptable for a hospital robot to collide with patients, doctors, and medical equipment on a regular basis.

To address this problem, iRobot has developed Ava™, a semiautonomous telepresence robot that includes capabilities for autonomous navigation, mapping, and obstacle avoidance (Figure 5.5). Ava is equipped with a planar lidar (light detection and ranging) sensor and PrimeSense 3D range sensors, similar to the one used in the Microsoft Kinect peripheral for the Xbox 360 and PCs. These sensors allow Ava to build a map of the environment, plan paths to user-specified destinations, and execute those paths while avoiding both stationary and moving obstacles. Ava's SLAM capability allows it to localize itself from a map at the same time that it is building that map. Ava uses an Android tablet, such as a Samsung Galaxy Tab or an iPad, as its head, providing two-way video communication between the remote operator and the people around the robot. Local users can also give instructions to Ava via the tablet's touchscreen. For example, Ava will display a map of the environment on the tablet, and the user can tap on any location to tell Ava to navigate to that location.

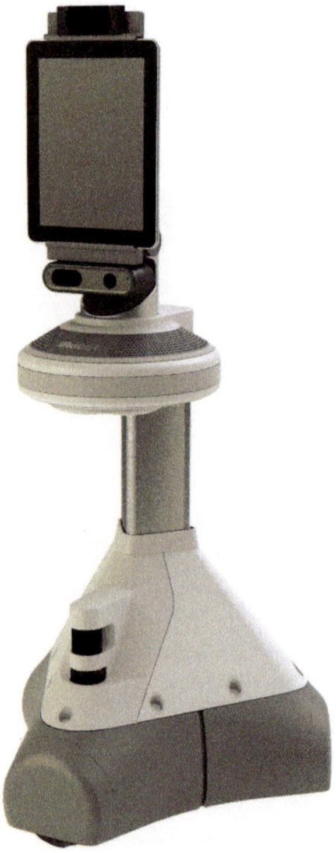

Figure 5.5 iRobot Ava™ telepresence robot. (courtesy of iRobot)

In the future, manipulator arms will allow telepresence robots to pick up, examine, move, and deliver objects. They will also be able to use tools and perform tasks requiring high levels of dexterity. For example, a service station might have a telepresence robot that allows expert mechanics located at central facilities to inspect and repair automobiles in the shop.

Increased Internet bandwidth will allow for more immersive telepresence through the use of head-mounted displays and tactile and haptic feedback. Recreational telepresence could provide people with experiences they might otherwise never be able to enjoy, due to expense, danger, or physical disability. For example, recreational telepresence robots could allow people to go walking in active volcano craters in Hawaii, scuba diving off the Great Barrier Reef via an unmanned underwater vehicle (UUV), or BASE (buildings, antennas, spans, and earth) jumping from mountains in Norway with a teleoperated glider.

5.6 Assistive Robots for the Elderly and Handicapped

Throughout the developed world, across the United States, Europe, and Japan, the proportion of residents over the age of 65 is growing rapidly. Caring for the elderly will become an increasing challenge in terms of cost and the number of available caregivers. Nursing home care is particularly expensive, and many elderly people would strongly prefer to live independently.

Telepresence robots will allow caregivers to check in on elderly relatives. Autonomous robots with manipulation capabilities will act as personal assistants, helping the elderly to lift heavy objects and reach high shelves. Assistant robots, transported by autonomous vehicles, will shop at supermarkets, deliver groceries to homes, and store items in the cupboard and refrigerator. These robots will also assist handicapped individuals who are confined to a wheelchair.

Advanced prosthetic technology will provide improved robot limbs for amputees. Sophisticated robot arms such as those developed for the DARPA Revolutionizing Prosthetics Program and robot legs such as the iWalk BiOM are increasingly able to duplicate the capabilities of human arms and legs. Patients will be able to control these prosthetics with their motor nerve impulses using EMGs (electromyographs) and, eventually, with their thoughts using EEGs (electroencephalographs). Currently, the signals detected using EEGs and EMGs are limited in terms of both precision and reliability. However, these limitations may be addressed in the future through more sophisticated sensor processing algorithms. Another alternative is to use electrodes known as ECoGs (electrocorticographs), which are implanted under the skull and make direct contact with the brain. ECoGs are able to monitor brain activity with far greater precision and reliability than noninvasive EEGs.

5.7 Humanoid Robots

And what of Rosey and other humanoid or near-humanoid robots?

Humanoid robots have a number of advantages and two key disadvantages compared with more specialized robot forms. One major advantage is that the humanoid form is general-purpose and can accomplish many different tasks. Like the character Neo in *The Matrix*, a humanoid robot could gain the ability to assemble furniture, fix a sink, iron clothes, or fire a rifle by simply downloading the appropriate program. A second advantage is that humanoid robots can make use of tools that were designed for humans. A third advantage is that man-made environments are typically designed for humans, and a humanoid robot could go anywhere that a human could go. A humanoid robot could walk through doorways, climb up and down stairs, and ride in vehicles. A fourth advantage is that the human form may be easier for people to relate to and interact with. In the future, humanoid robots may be able to use gestures to enhance spoken communication and use cues such as gaze direction to identify objects that are mentioned in conversation.

Unfortunately, humanoid robots suffer from two major disadvantages that limit their near-term use. The first is hardware cost. Any humanoid robot that duplicates all of the relevant degrees-of-freedom of the human body will be very complex and very expensive. Arms and hands that closely simulate human kinematics can carry six-digit price tags. When humanlike legs and sophisticated sensors for vision, tactile sensing, and force-based control are added, the price of a fully functional humanoid can reach millions of dollars. The second limitation is control complexity. Although advances have been made in controlling humanoid robots, like Honda's Asimo, and other legged robots, like Boston Dynamics's Big Dog, current robot control systems are far from being able to duplicate the manual dexterity or robust locomotion of a human being.

However, humanoid robot hardware and control software remain active areas of robotics research. DARPA has recently announced a Robotics Challenge for semiautonomous humanoid robots that can assist in disaster operations. The goal of this competition is to develop humanoid robots that can move through the rubble-strewn interior of a damaged building, climb a ladder, cut through a wall, and close a valve that is leaking gas. DARPA's Grand Challenge and Urban Challenge led to the development of autonomous vehicle technology that will likely make its way into consumer automobiles. Likewise, the new Robotics Challenge may contribute to the eventual development of practical humanoid robots.

An affordable, near-humanoid robot maid like Rosey is probably decades away. Sooner or later, though, it will almost certainly happen. Despite all of the challenges, the advantages of a general-purpose humanoid robot are too large to ignore or abandon.

5.8 The Next Ten Years

Over the next ten years, special-purpose consumer robots will become as commonplace as PCs and smartphones. Most homes will have multiple robots to perform cleaning and lawn care tasks. Robot vacuum cleaners, floor washers, lawn mowers, and pool cleaners will be joined by bathroom-cleaning and laundry-handling robots. Like dishwashers and microwave ovens, cleaning robots will become essential parts of the typical middle-class lifestyle, allowing consumers to have more free time to spend enjoying life.

Telepresence robots will become more common in many aspects of life. Health care robots will allow medical specialists to examine patients and consult on diagnoses at remote locations. Elder care robots will allow people to visit distant relatives and check on their well-being remotely. These robots will be equipped with manipulation capabilities to help the elderly and the handicapped perform everyday tasks. Advanced prosthetics will improve the quality of life for amputees by giving them more humanlike arm and leg replacements. Recreational telepresence will provide people with the ability to virtually travel to exotic locations and safely participate in extreme sports and other daredevil activities.

5.9 2020 and Beyond

Looking farther into the 21st century, greater autonomy will enable robots to perform increasingly complex tasks, and multirobot coordination techniques will allow groups of robots to work together. Autonomous vehicles will allow robots to travel to different locations to deliver products and perform services. Swarms of simple, insectlike robots will be able to collectively perform tasks such as cleaning or lawn maintenance.

Improvements in natural language understanding and gesture recognition will allow people to interact with robots in a more intuitive way, as if they were talking to a person rather than operating a machine. A person will be able to point at an object and say, "Put that in the closest, and then tidy up around here," and the robot will be able to follow these directions.

Brain-computer interfaces will allow amputees to control prosthetic limbs using their thoughts, and these limbs will provide humanlike strength and dexterity. Eventually, technology will offer prosthetic limbs and other artificial organs with greater-than-human capabilities, and some healthy people may choose to have their biological limbs and organs replaced by better, stronger, faster mechanical substitutes.

Like computers, electronics, and telecommunications in 2014, robotics will eventually affect every aspect of the human experience. Just as today's children are growing up with computers, smartphones, video games, and social networking, tomorrow's children will grow up with robot toys and friends around the world that they visit via telepresence. Instead of reading about dolphins, children will be able to swim alongside dolphins via semiautonomous UUVs.

At the other end of life, increasing amounts of elder care will be performed by autonomous and semiautonomous robots assisting or replacing traditional caregivers. The human lifespan will be extended as robotic devices are able to replace an increasing proportion of the human body. Many children born today will live to see the 22nd century, maintaining their health and vitality into their eighties and beyond thanks to robotic augmentation.

Humanoid robots will finally arrive in a form that is useful and affordable. These robots will act as general-purpose assistants that will adapt to perform a great variety of tasks and learn from experience. People name their Roomba robots, and soldiers have named and formed emotional attachments to their PackBot robots. These trends are bound to increase as robots become more intelligent and more autonomous. Eventually, people will see humanoid robots as companions or pets, rather than just machines, and will start treating them like members of the family—just like Rosey.

5.10 Conclusion

In the 21st century, consumer robots will expand from their current niches, such as floor cleaning, to touch every aspect of everyday life. Robots will maintain homes, teach children, and care for the elderly. By 2020, robots will become as ubiquitous as computers and smartphones are today. Telepresence will allow people to visit faraway places and physically interact with remote environments. Autonomous robots will advance from their current, insectlike levels of intelligence with the ability to communicate using natural language, learn from experience, and adapt to new situations. The robot revolution starts now.

5.11 References

5-1. Brooks, R. (1986). "A Robust Layered Control System for a Mobile Robot." *IEEE Journal of Robotics and Automation* 2 (1): pp. 14-23.

5-2. Arkin, R. (1998). *Behavior-Based Robotics.* MIT Press: Cambridge, MA.

5-3. Beer, R.D., Quinn, R.D., Chiel, H.J., and Ritzmann, R.E. 1997. "Biologically Inspired Approaches to Robotics." *Communications of the ACM* 40 (3): pp. 30-38.

5-4. Durrant-Whyte, H. (2006). "Simultaneous Localization and Mapping: Part I." *IEEE Robotics and Automation Magazine* 13 (2): pp. 99-110.

5-5. Felzenszwalb, P., Girshick, R., McAllester, D., and Ramanan, D. 2010. "Object Detection with Discriminatively Trained Part Based Models." *IEEE Transactions on Pattern Analysis and Machine Intelligence,* 32(9): pp. 1627-1645.

5-6. Yamauchi, B., and Massey, K. 2009. "Stingray: High-Speed Control of Small UGVs in Urban Terrain." *Proceedings of SPIE Vol. 7332: Unmanned Systems Technology XI,* pp. 73321T1-73321T12.

5-7. Glazer, E. (2012). "When a Robot Becomes the Life of the Party." *Wall Street Journal,* March 10, 2012, p. A1.

Chapter 6
Landscaping Applications

Alex D. Foessel-Bunting, John Deere Co.
Justin A. Craft, John Deere Co.

> The future belongs to those who see possibilities before they become obvious.
>
> John Scully

6.1 Chapter Summary

Landscaping applications of autonomous vehicles present challenges of outdoor operations (varying lighting and weather conditions, presence of people and property) and at the same time the benefit of semistructured areas (predictable driving paths, operation areas, tasks). Population and business trends are generating demand for increased work quality and output, while time and resources become scarcer. Technologies have matured and have demonstrated successful examples of autonomous vehicles executing some key tasks, such as mowing. We believe that this application area presents commercial opportunities, and remaining questions reside mostly in the business rather than in the technology aspect, with the exception of a key enabling technology: local positioning. Throughout, this chapter uses autonomous mowing to present business and technology considerations and requirements and to describe selected applications with varying degrees of commercial maturity, and it discusses the challenges of the next 20 years—mostly in bringing existing technologies to market in a profitable manner.

6.2 Introduction

Landscaping refers to activities that modify the visible features of an area of land, including living organisms such as flora or fauna, or what is commonly referred to as gardening, the art and craft of growing plants with a goal of creating a beautiful environment within the landscape; natural elements such as landforms, terrain shape, and elevation, or bodies of water; human elements such as structures, buildings, fences, or other material objects created and/or installed by humans; and abstract elements, such as the weather and lighting conditions.

Examples of landscapes applicable to this chapter are home gardens, sport fields, golf courses, commercial gardens, parks, highways, airports, and other areas where the landscape fulfills a very specific function. Landscaping aspects related to agricultural and mining activities are excluded from this chapter, as they are treated in other chapters in this book. Similarly, earth-moving aspects of landscape construction are also treated in other chapters.

When discussing the benefits of landscaping automation, it helps to divide the customers into two major segments, since they have different general requirements: the consumers and the commercial customers. In the following sections, the discussion often refers explicitly to these two customer segments.

In terms of the applications of landscaping, there is also a long list of possibilities such as grass mowing, trimming, irrigating, marking patterns and designs on the turf, cleaning, marking field lines, planting trees, fertilizing, and shaping. Because mowing exhibits the most progress in the adoption of autonomous solutions, both for homeowner gardens as well as for commercial applications (e.g., sports fields), this chapter will focus on autonomous lawnmowers, which should provide a good example of technology adoption.

6.2.1 Autonomous Lawn Mowing

Autonomous mowing competes with today's established technology of conventional push and rotary mowers. A look back in time on how the lawn mower was established shows decreased time between technical advancements with a relative overall low number of new adopted technologies. This can be seen in Figure 6.1.

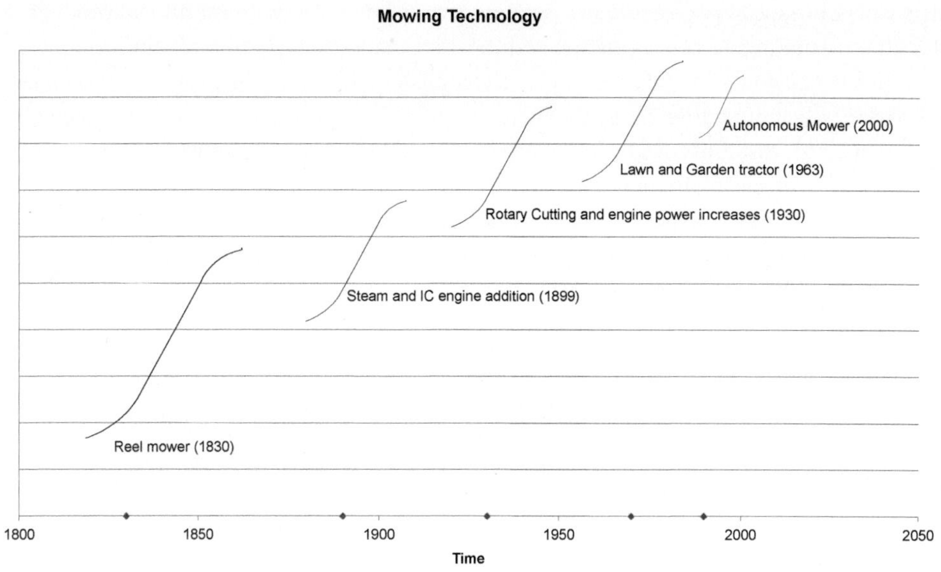

Figure 6.1 Technology innovation in mowing.

Current rotary blade cutting mowers have been in the industry since the 1930s and have an established following and cultural acceptance on how people maintain their yards. They are reliable, simple to operate, and have well-defined and understood interfaces with operators and supporting systems. Walk-behind and ride-on mowers are readily available and don't need additional property modification to begin use. However, these machines still require labor to operate and they cause fatigue to the user. Figure 6.1 shows that technology advancements are reducing the work the operator needs to expend for completing the task. This is done through adding more reliable power sources and eliminating operator walking.

6.2.2 Trends Driving Technology Adoption

The evolution of the lawnmower has been slow since the early 19th century. Since the first lawn mower was invented circa 1830, the evolution of the domain has continually reduced the human effort and work needed to complete the task. However, the reduction of manual effort or labor has not been the only characteristic of the domain. Improvements in cutting methods and manufacturing processes are other notable areas of advancement. Autonomous mowing is changing the business landscape due to the complexity of the system and at the rate of which the technology advances. These issues change how businesses must make decisions in the increasingly difficult product ecosystem. Previous history has shown that the lawn mowing domain has had slow advancements in adopting technology and subsequent diffusion. Yet today, autonomous mowing is an emerging technology in a slow-moving domain that is likely to provide first movers advantage.

Let's discuss this adoption first in the context of a homeowner, who wants to enjoy the look and feel of well-kept turf without the effort required by traditional means of maintenance. Autonomous mowing has been targeted toward the upper-income segments of the population due the initial high relative cost of the product. The current primary market exists where the social and regulatory demands make lawn care with traditional products a difficult undertaking. Noise regulation and the disturbance that traditional mowers create constrains the customer on when they can accomplish their lawn care. Restrictions on user's time and increasingly busier time schedules compound the problem in addition to the noise regulations. Age, disability, and allergic conditions also limit the capability or willingness of homeowners to engaging in mowing. An autonomous mower creates value by allowing the lawn care to occur with respect to ever-increasing constraints. A quiet product that runs on electricity and that can mow without intervention eases these constraints. Thus, these advances allow the consumer to enjoy the benefits of a well-kept lawn.

This context should also be expanded to a parallel view of a company that needs to maintain a turf field at the lowest effort and cost (i.e. corporations and sports complexes). Due to sports fields being designed in predictable sizes and shapes, the complexity of a worksite is lower compared with the complexity of the countless different shapes of home gardens. Even though these structures create easier navigation, the scheduling portion of a sports field is increasingly complex. An autonomous mower is required to completely mow a field between games and training sessions. Therefore, scheduling complexity is elevated due to decreased availability to mow. Increased efficiency and mowing power are required to complete jobs in the short time requirements between games. This scheduling difficulty drives higher product complexity and performance in contrast with a consumer autonomous mower that normally completes a mowing job over the course of days. This reduced consumer requirement allows the manufacturer to keep the complexity and cost of the product low.

6.2.3 Value Proposition–Homeowner Customer

In the case of the homeowner, a key driver is the desire to enjoy the lawn while reducing the effort to maintain it. Enjoyment of the lawn can have different meanings, since some homeowners like to spend time with family and friends in activities such as cooking, playing, or just talking, whereas others prefer to spend time working in their garden. In this last group, often mowing is considered the least-rewarding activity, as compared with planting, trimming, or harvesting fruits and vegetables from the same garden.

In addition to the restrictions mentioned earlier (time scarcity, noise regulations, and lack of capacity to mow), traditional mowing, executed occasionally, creates clippings. The disposal of this organic matter becomes an additional burden. Cities charge for disposal of grass clipping bags, and cleaning the car after the transport adds to the effort and time related to mowing.

In summary, homeowners prefer to enjoy the garden without the time or hassle involved in the mowing aspect of it, which is becoming increasingly burdensome as time availability decreases, and the hassle makes it a less rewarding activity.

6.2.4 Value Proposition–Commercial Customer

An example of a commercial customer could be a grounds maintenance crew of a professional sports stadium. The expectations are high: players expect the field ready and available to practice and play; fans, TV crews, and companies contracting advertisement expect the field to look sharp from all seats as well as captured by the TV cameras. Cut quality and line straightness is a key factor. There is pressure on the crew to deliver the field in a short time for the upcoming practices and for the "big game."

An autonomous machine that can deliver a well-mowed field can not only bring better output quality, but frees the time to work on other activities. The crew is able to do other productive work while a mower is mowing. This is a major deal because it changes the whole "time to mow" equation. In a sense, it does not matter how long it takes to mow (within reason) because other processes were getting done simultaneously. This, by itself, could actually be enough to justify the value of the mower. There is another aspect that brings additional value. For example when there is scarcity of resources (think vacations, lack of temporary staff, and so on), the availability of getting one big task out of the way allows the reduced team to focus on other tasks and still get the job done in face of otherwise trying circumstances.

In summary, automation doesn't necessarily eliminate labor, but it does change the timing of the day and reduces the stress level of the job. More importantly, automation of turf maintenance brings workers lower stress, better results, and the ability to focus more energy on other important aspects.

6.3 Requirements and Technologies

To understand the choices and integration of technologies to enable autonomous operation, it is necessary to understand the environmental requirements, schedule and workflow integration, and functional requirements associated with autonomous mowing—and by extension, to other autonomous vehicle applications in landscape maintenance.

6.3.1 Environmental Requirements

Landscape maintenance is an outdoor activity, where the environment is subject to weather, has a diversity of objects and obstacles, and sky-obstructing features, such as trees and buildings. Also, since landscapes are used for recreation, people are often present. These characteristics are important because they drive specific requirements to the technologies used in automation. For example, sky obstructions challenge GPS-based positioning. Presence of people and objects makes safeguarding a key requirement, more so than in other applications where people are seldom in the area or not present at all (e.g., mining operations or planetary exploration).

6.3.2 Scheduling and Workflow

Although the focus of autonomous vehicle development is traditionally on the unmanned execution of work tasks, the scheduling of the execution and synchronization with the workflow is critical for success. For example, in the case of a homeowner application, the scheduling requires consideration of the family lifestyle. When the family likes to spend afternoons enjoying the garden, a late night or morning maintenance schedule might be a better fit. In addition, an autonomous mower should allow for an easy method to interrupt the operation in case there is an unplanned occasion in the garden. Similarly, mowing a stadium or golf course requires careful scheduling to avoid interrupting or preventing a game or training session. In addition, synchronization of the mowing with irrigation schedules is also important to avoid driving on wet turf, which can cause undue wear and tear.

6.3.3 Functional Requirements

An autonomous mowing in a home garden, sports turf, or commercial landscape requires the ability to perform several functions. They are listed next to guide the selection of technologies, which are described thereafter.

6.3.3.1 Containment, Mowing Area, and Exclusions

For safety and correct work execution, an autonomous machine must be kept contained in its work area. The definition and identification of the driving limits as well as the definition of the mowing area are critical for good mowing results. In addition, any exceptions to mowing or driving, such as flowerbeds in the case of a garden, or the dirt areas, in case of sports fields, must be avoided. Regardless of the application, respecting the borders of driving and mowing requires high accuracy.

6.3.3.2 Area Coverage and Coverage Approaches

Mowing is an activity executed over a patch or region of terrain (as opposed to transportation, which happens along a line, such as a road). Coverage completeness is important for a quality result in most applications (highway driving may be an exception). In addition to completeness, in some applications it matters how the area is covered. An example is a baseball field, where the effect of driving straight lines in opposite directions or in a checkerboard path is necessary to create the visual pattern effect that distinguishes professionally maintained fields. In other cases, such as garden, even-looking grass, regardless of how it has been mowed, matters more.

6.3.3.3 Recharging

An autonomous vehicle, in addition to executing its work, requires frequent refueling or recharging. In the case of autonomous mowing, most of the commercially available products are electric. For continued operation, the ability to drive to a recharging station and establish an electrical connection is required. In the case of combustion-based machines, the ability to drive to a refueling station and execute the fuel transfer would be necessary. In both cases, driving to the station requires ability to locate, drive, and maneuver appropriately.

6.3.3.4 Safeguarding People, Property, and Turf

As described earlier, an autonomous machine supporting landscaping maintenance is expected to operate in a cluttered environment with frequent human presence and possibly additional machinery. The protection of life and property must be a key element of the machine function, for which specific sensing modalities and risk mitigation should be considered.

In addition to safeguarding people and property, protecting the turf is important. There is little value in a machine that mows perfectly, yet at the same time wears the turf with a poor end appearance. This aspect is particularly important when there is no human surveillance of the operation, which could quickly interrupt a damaging maneuver or operation.

6.3.4 Positioning Technologies

To meet the requirements of containment, mowing area definition, and exclusion, as well as to assist with the location of the charging station, there are several approaches. The most traditional method in autonomous mowing is to use a buried wire.

There are many approaches to local positioning of the vehicle. However, there are primary technologies emerging within the robotics applications. The adoption of a particular technology is to be evaluated through analyzing the "S" curve trajectory in relationship to the performance/cost ratio needed for the planned business strategy. Through this method, the proper technology can be selected for a product platform and business strategy. A brief description of the current key and primary local positioning strategies follows.

6.3.4.1 Boundary Wires

The most common technology in autonomous mowing is the use of a wire that is run along the perimeter of the customer's property. This wire can be laid on top of the grass or buried up to six inches deep below the surface. In conjunction with the installed wire, a base station is used to transmit an electromagnetic signal to the autonomous mower through that wire. The mower uses this signal to indicate if the mower is in or out of bounds relative to the installed perimeter. Wire technologies also allow approximations of distance from the perimeter, providing aid in certain navigation decisions. Due to the limited information regarding local position within the yard, this technology generally is relegated to random navigation. It is in the search for increased efficiency from accurate local positioning that the following technologies have the goal of path planning to develop the next dominant machine form and function.

6.3.4.2 Simultaneous Localization and Mapping

SLAM (simultaneous localization and mapping) methods use mapping of the environment through feature observations and state estimates of the dynamics of the robot motion. The problem is formulated as an optimization problem where the objective is the most probable path for success, given the constraints of obstacle identification from the sensors and dynamics of the vehicle. The latest method deployed in the paper uses stochastic gradient descent and incremental state representation for optimization algorithm formulation [6-1]. This is a nonlinear method that reduces the computational power over traditional linearization techniques. One such linearization technique is the extended Kalman filter, which lowers the necessary computational power by only focusing on the most relevant portions of the map.

6.3.4.3 Vision-based Positioning

Optical-based sensors have become, in the realm of robotics, the most used and developed technology to sense the robots' surroundings. The main types of optical sensors are charge-coupled device stereo cameras or laser scanning devices that build maps of the surrounding terrain or obstacles. Both of these methods have limitations. These limitations are best described in findings from autonomous indoor aerial vehicles applications. This realm of robotics is a lead user of the technologies and extract critical limitations in trying to achieve their goal of flying autonomously in areas where GPS is not available. Laser range finding is useful when the objects in the workspace are unique and not homogeneous, and since it formats the data into 2D slices it does not perceive objects outside the tuned range [6-2]. In contrast, stereo vision cameras provide large quantities of 3D data but have difficulty distinguishing unique items in the field of view. Stereo cameras have a limited angle of view and become computationally expensive to utilize the data generated. However, research and application of optical-based sensors are continuing to grow due to the functionality of operating indoors and the increased accuracy over GPS.

6.3.4.4 Global Positioning System

Global Positioning Systems have proved their useful functionality in providing location information data in such products as cell phones and car navigation units. These products lend an important comparison for the application to that of autonomous mowers in that they are commercial products that compete heavily on cost for increased adoption. Comparing the accuracy of these budget-sensitive systems gives insights to the performance limitations of autonomous mowing. In contrast to the optical-based technologies, the GPS systems are only functional in locations with a line of sight to the sky. Indoor applications are not suitable for GPS systems as no data is available from the satellites without the line of site. The primary drawback in using GPS for local positioning is the overall accuracy of the system versus the cost.

To study the accuracy of GPS systems, it is currently accepted to break the grade of the system into three levels [6-3].

Level	Accuracy	Cost (USD)
1	1cm	25000
2	2-5cm	2000-12000
3	5-10cm	100

Since level 3 GPS is affordable to use in an autonomous mower, the accuracy of the system is important to determine the capabilities for potential path planning and mowing efficiency gains. However, the level of inaccuracy for level 3 is not acceptable for functional path planning. These levels are much greater than the typical width of 0.5 m for a vehicle and indicate that the vehicle's location is not known within an order of magnitude of the vehicle's size. This lack of precision makes path planning a useless endeavor with GPS. At 10 meters, the vehicle would be trampling important items such as flower gardens and mowing far outside the owner's property.

However, compared with the current products that randomly navigate the yard, GPS enables course map generation to create relatively complete coverage of the work space. Instead of complete path planning, the vehicle can navigate to different areas of the yard that have not been addressed in reasonable timeframes. In difficult corner conditions, the course map from the GPS saves valuable coverage time over the current techniques. Although these efficiency gains are not ideal compared with complete path planning, the benefits are substantial in large areas. In large yard segments, the course maps save significant portions of time and minimize the need for excessively large batteries.

6.3.4.5 Landmark-based Positioning

Distributed network systems are utilized in areas where the position of nodes is estimated through distance measurements. This is an algorithm that can be utilized in noisy sensor environments such as ultrasonic, radio frequency, or infrared sensor technologies. The application has been utilized in mapping the working area of a mobile robot using ultrasonic sensors [6-4]. Instead of using a high-cost GPS, laser, or camera vision system, a low-cost sensor network such as radio frequency identification could be utilized to map the workspace. Combining a robust network localization algorithm and low-cost passive sensors distributed within the workspace gives a viable means of creating a form of path planning and increasingly efficient navigation.

6.3.5 Technology Scouting Opportunities

At the same time that companies invest in product development, other organizations promote technology development around autonomous mowing. University research is a primary means to identify new areas of emerging technologies in applications such as the AUVSI competitions for unmanned vehicles [6-5] and ION competition for autonomous mowing.

The Institute of Navigation (ION) has sponsored an autonomous contest for eight years. The purpose of this competition is to challenge students—our future engineers, scientists, and problems solvers—to design and operate a robotic unmanned lawn mower using the art and science of navigation to rapidly and accurately mow a field of grass. The competition provides an excellent challenge for students completing their engineering studies and working on senior projects.

Pushing limits in a different way is the Association for Unmanned Vehicle Systems International (AUVSI). This organization has yearly competitions that push universities to develop the next level of automata. These competitions are difficult because of the challenges given and the conflicting requirements of achieving those stated goals. Due to the fast dynamics of the aerial vehicles and the need for small, light, and low-power computation, it pushes the envelope of technology. This is a good place to look for the next possible disruptive technology.

While these competitions push the boundaries of technology, there is still a significant effort in technology selection for product design. The technology eventually transfers to consumer products when the cost and manufacturing feasibility is viable. Estimation of further development cost is a key factor when deciding to implement state-of-the-art research technologies.

6.4 Selected Autonomous Landscaping Applications

6.4.1 Lawn Mowing Home Gardens

There are several manufacturers of autonomous mowers for homeowner application, such as Husqvarna, John Deere, Friendly Robotics, Zuchetti, Bosch, and Worx. These machines are also sold under various other brands. The models cover gardens from 500 sqm up to 3500 sqm.

There is a wide difference between traditional mowing and autonomous mowing. The principle of traditional mowing is to mow periodically, but with longer intervals from one mowing to the next. This is effectively getting long grass short. Conversely, autonomous mowing takes place frequently, even daily, so the concept is of maintaining short grass short. Because of frequent mowing, the mower also can operate in wet conditions. This is achievable because the short grass clippings don´t clump when left on the ground. It is generally understood that this approach improves the health of the turf, which can be explained with the increased resulting organic fertilization.

Current state of the art for consumer autonomous mower operation requires a boundary wire for delivery operation containment, to define the mowing area, to keep out of areas, and to indicate the path to the charging station. This boundary wire can be installed as buried or fixed with stakes to the garden surface. In some instances, such as the AutoMow from Husqvarna or Tango (Figure 6.2) from John Deere, there are additional wires to provide a shorter path to the charging station or to allow for precise docking. Multiple strategies such as the use of spiral patterns, distance-defined launch points along the perimeter wire, and perimeter clean-up laps help to improve the coverage efficiency. Zuchetti, an Italian manufacturer of autonomous mowers, also has a containment approach based on a grass sensor, which detects the boundary between grass and nongrass areas, to determine where to mow and drive.

Safeguarding is most typically achieved with the combination of limited speed and collision detection. In some cases, a floating shell and force sensors are used to detect a collision, whereas other designs use an air pressure sensor that senses pressure changes in a front-mounted bumper. Other designs use high torque sensing at the driving wheels. More-sophisticated designs include ultrasonic sensors (similar to the ones used for parking assistance in contemporary cars) to slow down the speed of the mower and limit the impulse of the machine before a collision. The turf care is achieved by controlling maximum torque and limiting the transit along the wire.

The market for consumer autonomous mowers keeps growing fast, approaching more than 100,000 units sold per year in 2011, mostly in Europe. This number is still small compared with the several million walk-behind mowers sold in the same market, Yet if the market development of autonomous vacuum cleaners is any indication of the adoption of

Figure 6.2 Tango, an example of a consumer autonomous mower. (courtesy of John Deere)

technologies to save time and effort, we should expect increasing replacement of the traditional solution.

6.4.2 Mowing Large Turf and Sports Fields

Application of the random autonomous mowers is finding its way to larger areas. For example, Belrobotics, a Belgium-based producer of automatic robot mowers for large grounds, sells the BigMow (Figure 6.3)—the largest random autonomous mower at present—which covers lawns up to 2.5 ha. A new robot to be launched in 2014 will cover areas up to 8 ha. Belrobotics sells its equipment in Western Europe through a network of service importers and dealers and to original equipment manufacturer partners.

- Football pitches of all categories, from leading international clubs to amateur clubs all over Europe
- Large turfs around factories, office blocks, hospitals, hotels, country clubs, councils, military zones, and so forth
- Large gardens and parks
- Golf practices (driving ranges, approach and putting) where Belrobotics robots concurrently collect golf balls and mow the practice turf
- Industrial applications such as grass farms, photovoltaic fields, high-voltage and power stations

Figure 6.3 BigMow maintaining a large commercial ground. (courtesy of Belrobotics)

6.4.3 Precise Stadium Mowing

While random mowing technology is making inroads in the sports turf mowing market, other companies are experimenting with more-sophisticated approaches. Toro and John Deere have presented prototypes of autonomous mowers using existing fairway mowing platforms. These prototypes integrate state-of-the-art perception and local position technologies, which enable very precise control of the platform and mowing reels. Instead of the random path followed by the machines described here, these prototypes navigate using precise local positioning systems (with accuracy of 1–2 cm, comparable with real-time kinetic GPS) based on radio frequency or laser systems, and use deterministic paths that mimic the driving of expert operators. One example is the Toro stadium mower developed in conjunction with Carnegie Mellon University [6-6], demonstrated in sports fields and golf fairways (Figure 6.4). Another is the autonomous stadium mower effort by John Deere, which deployed prototypes in three stadiums in Los Angeles, Phoenix, and Durham to capture valuable insights in the customer requirements and technology maturity [6-7].

Figure 6.4 Prototype of autonomous stadium mower deployed in soccer field.

6.4.4 Golf Course Maintenance

Precise Path Robotics, Inc., manufacturer of the RG3, has brought to market the industry's first and only robotic greens mower specifically designed to reduce operating costs while simultaneously improving the course quality conditions (Figure 6.5). The RG3 has been proven on golf courses across the country to make the golf course maintenance staff 50% more productive during the critical course preparation hours every morning while providing excellent consistency across all greens.

Figure 6.5 Precise Path autonomous greens mower. (courtesy of Precise Path Robotics)

In addition to labor efficiencies, the RG3 provides consistent putting surface across greens by combining rolling and mowing in the same unit and eliminating the influence of an operator. Additional benefits of the RG3's include these:

- Exact clean-up cut that eliminates scalping and greens creep over time due to operator error
- Proven greens speed increases (independent studies show up to 12 inches of additional ball roll)
- Nighttime mowing capability
- Controlled, turf-friendly turns

Key technologies that deliver this required performance are an ultrasonic-based localization system, a Li-Ion battery, and precise controls for mowing and maneuvering. The localization system uses beacons, which once placed around the greens provide a reference for the mower to execute the functions of containment and precise area coverage.

6.4.5 Advertisement and Art on the Turf

In addition to the checkerboard patterns common in professional stadiums, occasionally fans and TV viewers are treated with logos or other figures visible on the turf. These grass-bent, turf logos and images produced can create high-impact turf billboards at sporting events [6-8]. The art of grass bending consists of redirecting blades of grass so that light reflections produce dramatic, visual effects. Rollers and brushes are programmed to create intricate, elaborate grass designs. Manual implementation of such designs on a sports stadium takes time and is incompatible with a game. Start-ups such as Gentry Robotics [6-9] attempt to use autonomous vehicles to deliver these images on the turf in real time. The speed, efficiency, and degree of detail that can be delivered with an automated vehicle carrying a grass-bending tool could not be matched manually by the best of turf specialists. When thousands and thousands of blades of grass are bent in contrasting directions, a pattern emerges, a pattern that can be spectacular.

6.5 Commercialization Considerations

Robotic technology has seen varied progress toward market introduction. In some cases, such as with household appliances, the market is established and new products are surpassing the market volumes of existing or traditional solutions[1]. In many other cases, the solutions are far from market introduction and remain in the research phase. Regardless, at some point the development of landscaping automation solutions, when aimed toward commercialization, will face the definition of a commercialization strategy, including sales, distribution, and support. The following sections describe key considerations that may help in accelerating the market introduction: balance between a technology-focused and a service-based solution.

6.5.1 Balancing Technology and Service

The creation of an automated solution for landscaping has several challenges, such as understanding the requirements, evolving customer needs, continuous improvement of the technology, product updates (especially software), and a diversity of customer needs representing each customer segment. These challenges become even more difficult when the expectation is that the technology needs to be in place to address these aspects. Such a technology-driven approach would make the risk of the project unnecessarily high.

In some cases, it is possible to ease the requirements put on the technology and the support and service that can be provided by other players, such as service partners, dealers, and other stakeholders as part of the distribution and support. With such an approach, the product does not have to evolve as quickly and with such complexity to serve the diversity and dynamics of customer needs, but instead may rely on customization and configuration provided by service providers that are closer to each customer and can better provide the adaptation.

For example, in the case of a homeowner mowing solution, key benefits are freeing time and ease of mowing. They can be realized with a range of approaches, from hiring a professional landscaping company, which clearly provides these benefits (the customer signs a check and agrees on the schedule), to a fully automated solution that would be placed on a garden and would automatically recognize the areas to be mowed and adapt to the grass growth to set its own schedule. This would be a purely technical solution, yet it may be difficult, with today's technology, to achieve in a profitable manner. The current solution is a balance between the technology capabilities and the service provided by an installer, typically the dealer.

Autonomous mowing also faces competition from the landscaping service industry. Firms providing a service to complete landscaping tasks could provide a barrier to autonomous mowers securing their place in the consumer market. A known strength of a landscaping service is that it is a flexible business strategy, giving the consumer the value of a well-maintained property with no physical labor. Not having to purchase equipment is an attractive alternative for the consumer. However, the penetration of autonomous mowers as a primary tool of the service industry is a relative weakness of a service-only strategy. A landscaping service provides an alternative over owning and maintaining lawn equipment. Thus, a strong service industry creates a barrier of entry of the autonomous mower to the direct consumer. The consumer receives the same benefit of eliminating lawn maintenance by choosing a landscaping service. Additionally, a landscaping service can bundle additional services to increase value and the aesthetics of the consumer's property. An added dimension of giving specialized services such as chemical applications, aeration, and knowledge of agronomy allows a landscaping service to maximize the appearance of a yard. A service not only maintains the height of the grass but provides technical tools and know-how to keep a lawn looking at its finest. An autonomous mower could be adapted to provide other functionality, but most likely would struggle to provide the same value as a team of landscaping professionals. The adoption of robotic mowers as tools for the landscape professionals is an alternate avenue for diffusion. A reliable, efficient, and economical solution to manual labor is of interest to a landscaping service provider. The robotic mowers eliminate people working around dangerous machinery and allow them to work on more valuable tasks. It is in the context of this larger value system that autonomous mowing finds its competitive leverage.

Defining corporate home development companies as large-scale builders that develop large subdivisions or housing communities provides an opportunity to create a streamlined solution for the homeowner. Input into the beginning stages of creating the community architecture allows the system to be tailored to current levels of autonomous mowing technology. A corporate builder and autonomous mowing manufacturer can optimize the system to provide the best solution and value for the homeowner. The potential idea is to market autonomous lawn care as a feature of the community and plan the community's architecture to enable success with the current state of autonomous mowing technology. An interesting paradox to this strategy is that an optimized solution at this level could reduce the number of potential autonomous mowers sold but at the same time have greater success in providing value to the homeowner. Due to the planning, it is likely that fewer robotic mowers are needed, compared to the case where each homeowner buys their own.

Sports venues such as golf courses and baseball and football fields have an inherent need to maintain turf. The operators of these venues provide a service that is directly linked to the condition and appearance of the turf. A golf course's improved appearance and condition over rival courses allows the management to charge a premium for its use. A sports venue has turf maintenance as a fixed cost in providing a value to their customers. Gains in efficiency in this task would allow for higher provider profit margins. These venues have increased need for path planning due to their size and the demands for perfectly cut and striped turf.

6.6 2020 and Beyond: Commercial Opportunities

The evolution of the business domain for outdoor power products is currently at the epoch for autonomous mowing. A transition in this domain is under way to determine the relationship between the emerging autonomous mowing technology and that of the incumbent riding and push type lawnmowers. The ecosystem for this domain in past history has been dominated by a vertical industrial architecture [6-10]. Due to the increased complexity of autonomous mowing, success demands a focused and sustainable architecture for the product and the ecosystem. A company that nurtures a proper plan for building a business ecosystem, executes a robust architecture, and understands proper timing and placement in the domain has the best chance for capturing value from the evolution of autonomous mowing. Examination of the niches, current market ecosystem structure, and previous business strategies are analyzed to provide insight for creating a robust business strategy for autonomous mowing.

Autonomous mowing provides increased opportunity to create value in the business domain of outdoor power products. A broad definition of the ecosystem that an autonomous lawnmower operates in is the following: all things outside on the customer's property. The scope of the business ecosystem has been broadened to provide a more complete picture of the aspects needed to create and capture maximum value for the customer. The definition not only includes the yard but incorporates a view for a total solution that maintains the appearance of the whole property. For example, the exterior surface of the house or building is included in this definition and provides an additional area to create value in the ecosystem. This definition has a key component in that it changes the traditional context for outdoor power products and allows for a proper definition of the key niches. This high-level viewpoint creates a framework to analyze and deliver the most value to the customer

Critical niche categories are defined as the following: lawn maintenance, exterior building maintenance, and peripheral applications. These categories are a top-level view and have been consolidated for simplicity. The listed individual niche activities are not all-inclusive, however, and are likely to change over time. Table 6.1 shows representative individual activities that contribute to the top level categories.

Table 6.1 Individual Niche Activities

Lawn maintenance	Exterior building maintenance	Peripheral applications
Maintaining height of grass	Roof maintenance	Hedge/Bush maintenance
Fertilizer application	Window cleaning	Gardening functions
Targeted application of chemicals	Gutter cleaning	Tree trimming
Leaf collection	Painting	Pool cleaning
Pesticide application	Exterior washing	
Trimming around objects	Snow blowing	
Aeration		
Water Management		

Currently, a wide range of products are needed to provide the functionality in the table's listed activities. The listed niches contain a wide variety of functionality that an autonomous solution could span. A leader in this domain would have a strategy to maximize applications to create and capture additional value. Currently, a solution that captures value across all niches has not been attained. However, it can be seen in Figure 6.6 and Figure 6.7 that a potential path exists for an autonomous solution that spans all niches for the customer.

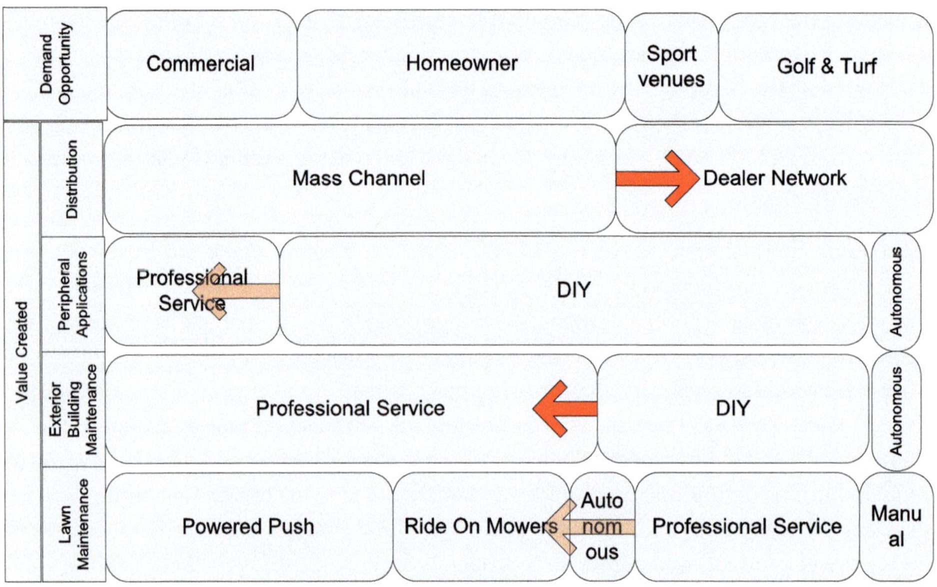

Figure 6.6 Business ecosystem segment trend.

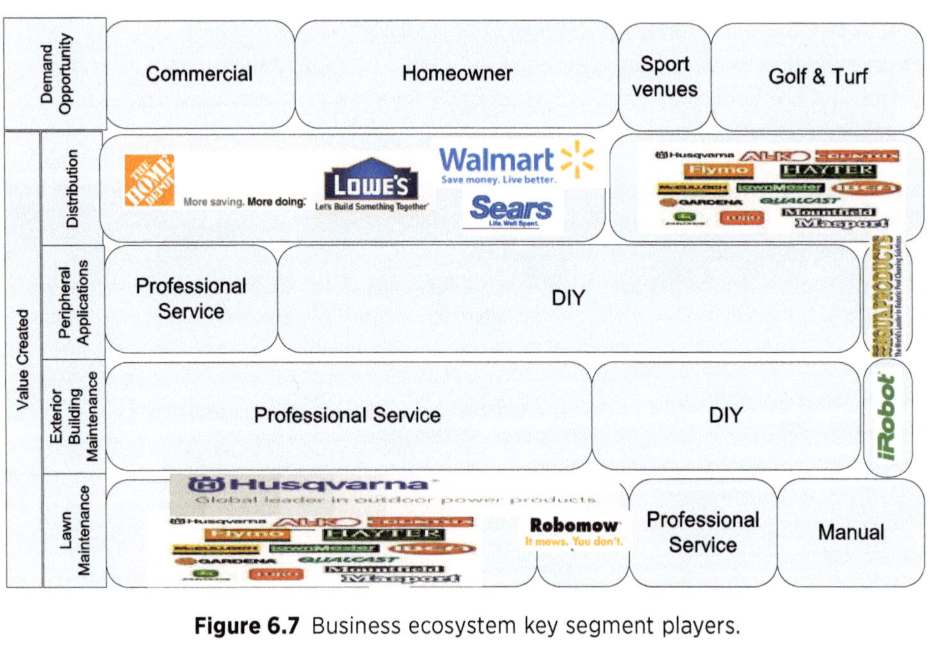

Figure 6.7 Business ecosystem key segment players.

Figure 6.6 and 6.7 also show that the business ecosystem players have all taken specialist roles in autonomous solutions as defined by this framework. Creating an autonomous segment allowed these companies to capture additional value and enjoy a leader role for their efforts. These companies captured value by creating the autonomous segment and being first movers for the execution of a first dominant design. However, these roles are likely to change in the future as the leaders plan architectures to capture additional segments of the ecosystem.

The future evolution of the outdoor power products domain is most likely shaped and controlled by a company that executes a keystone/leader [6-11] strategy for the execution of innovation. As seen in the business ecosystem maps, the current solution space is diverse and fragmented. The solution that spans and captures value across the defined niches requires technologies from a diverse set of disciplines. In order to shape the evolution of the domain, a company needs to create a modular architecture that brings together the diverse functionality and ecosystem players. A company that executes strategic use of architectural knowledge [6-12] can control and define the domain's evolution. By creating a robust architecture, a company can integrate players from across the niches and minimize their architectural "footprint" [6-12]. This strategy is critical for capturing maximum value from the fast-moving digital technologies that facilitate an autonomous solution.

An aspect that can be utilized for companies vying to become architectural leaders is to create and control the standards for the autonomous system. As the solution becomes modular, the ability to control key bottlenecks [6-12] in the architecture is critical to capturing and maintaining value. By defining standards, the company can control the value capture in the domain. In the battle for the next dominant design, the creation of standards defines compatible and incompatible technologies for the domain. This allows an open communication path for supporting technologies to be integrated and defined. In order to successfully complete a standards battle, a company must have several key assets. These assets are defined by Shapiro and Varian as follows: control over installed base of users, intellectual property rights, ability to innovate, first mover advantages, manufacturing capabilities, strength and complements, and brand name and reputation [6-13].

Current business models for the outdoor power products domain have focused on a particular niche segment. Autonomous solutions have only concentrated on performing one task, such as just maintaining the height of grass. Adopting these strategies, which have formerly controlled the domain, cannot capture the diverse and fast-changing value in the ecosystem. Previous strategies were bounded by the rationalization that innovative products only deliver value for a particular niche. Autonomous mowing has created a platform that is able to deliver additional value across multiple niches. A keystone/leader of the domain recognizes that even though autonomous mowing is in its birth stage of evolution, it can be transformed into a larger value proposition of autonomous household solutions during the expansion stage of evolution. A company that adopts this strategy grows the domain and the potential for capturing value. One such strategy is to take a horizontal approach to autonomous landscape solutions.

In adopting a horizontal industry architecture [6-10] strategy, a company has to focus on maintaining the value capture in the domain. Increasing modularity increases the chances for imitation and value erosion. Modularity shifts the locus of innovation to that of the supporting technologies, that is, a component or subsystem [6-10]. An applicable regime that increases the barrier to entry is to control the architecture of the total system [6-10]. As the total system architecture increases value capture across the niches, the barrier to entry is also increased. Additionally, over time increased path dependency is built, and switching costs become an additional barrier for imitation. Therefore, the locus of innovation is maintained to provide additional value for the customer in the ecosystem through architectural innovation. Understanding these dynamics is an important consideration during the leadership phase of the domains evolution.

6.7 Conclusions

The use of autonomous vehicles for tasks in landscaping and landscape maintenance is reaching a disruption opportunity. For example, university teams, research groups in companies, and groups of students in contests have demonstrated repeatedly successful autonomy in turf and lawn mowing. Basic random and containment technology enables homeowner mowing solutions that have reached hundreds of thousands of machines sold, mostly in Europe. Commercialization of precise machines for golf automation is ongoing in the United States. The remaining questions are not whether autonomous vehicles have an important role in this industry or when autonomous vehicle will profitably enter the market. The answers are yes and now for several companies. We hope the discussions in this chapter will help others to join this innovation journey as well.

6.8 References

6-1. Olson, E., Leonard, J., and Teller, S. (2007). *Spatially-Adaptive Learning Rates for Online Incremental SLAM.* Cambridge, MA: MIT Computer Science and Artificial Intelligence Laboratory.

6-2. Achtelik, M., Bachrach, A., He, R., Prentice, S., and Roy, N. (2009). Stereo Vision and Laser Odometry for Autonomous Helicopters in GPS-denied Indoor Environments. *Proceedings of the SPIE Unmanned Systems Technology XI.* Orlando, FL: Technische Universit¨at M¨unchen, Germany;Massachusetts Institute of Technology, Cambridge, MA.

6-3. Wing, M. G., Eklund, A., and Kellog, L.D. (2005 June). "Consumer-Grade Global Positioning System (GPS) Accuracy and Reliability." *Journal of Forestry* , p. 169.

6-4. Moore, D., Leonard, J., Rus, D., and Teller, S. (2004). Robust Distributed Network Localization. *Proceedings of the Second ACM Conference on Embedded Networked Sensor Systems (SenSys '04),* pp. 50-61. Baltimore, MD.

6-5. AUVSI. (2010). *Association for Unmanned Vehicle Systems International.* Retrieved 2010 15-11 from AUVSI: http://www.auvsi.org/auvsi/auvsi/Events/AUVSIStudentCompetitions/Default.aspx.

6-6. Roth, S.A., and Singh, S. (August 2004). "Application of Robust, High-Accuracy Positioning for Autonomous Ground Vehicles," AUVSI Unmanned Systems North America.

6-7. Zeitzew, M. (July 2007). "Autonomous Utility Mower." *Agricultural Engineering International: The CIGRE journal*. Manuscript ATOE 07 016. Vol. IX.

6-8. Mellor, D.R. (2001). *Picture Perfect: Mowing Techniques for Lawns, Landscapes, and Sports.* New York: Wiley.

6-9. Gentry 21st Century Systems, Inc. "Home Page." Accessed www.gentryrobotics.com. Accessed February 6, 2014.

6-10. Pisano, G., and Teece, D. (2007). "How to Capture Value from Innovation: Shaping Intellectual Property." *California Management Review.*

6-11. Iansiti, M., and Levien, R. (2004). "Strategy as Ecology." *Harvard Business Review.*

6-12. Baldwin, C., and Clark, K. (2006). "Architectual Innovation and Dynamic Competition: The Smaller "Footprint" Strategy." Harvard Business Review.

6-13. Shapiro, C., and Varian, H. (1999). "The Art of Standards Wars." *California Management Review* , Volume 41.

Chapter 7
Unmanned Maritime Vehicles: New Options for Ocean Operations

Justin Manley, Teledyne Marine Systems
Thomas Altshuler, Teledyne Marine Systems

How inappropriate to call this planet "Earth," when it is clearly "Ocean."

Arthur C. Clarke

7.1 Introduction: Unmanned Maritime Vehicles

Today we are used to "robots" and unmanned systems in many domains. The public has become familiar with "drones" flying over war zones and emergency actions such as wildfires. Land robots are immediately recognized for their roles in combat and police actions, such as the investigation of suspected bombs. But the diverse ecosystem of unmanned maritime vehicles (UMVs) is often overlooked. These systems have been under development for decades, but, as with many developments in ocean science and technology, the veil of seawater hid this from common view. As an ocean planet and a global economy connected by sea-lanes, there are many roles for UMVs of many sizes and shapes.

The UMV space offers a series of names and acronyms, the most common of which include remotely operated vehicle (ROV), autonomous underwater vehicle (AUV), unmanned undersea vehicle (UUV), unmanned surface vehicle (USV), autonomous surface vehicle (ASV), autonomous intervention vehicle (AIV). A few trends are immediately clear in these names. First, "vehicles" in this sector are not designed for carrying people or cargo; rather, they provide accesses for tools and sensors. In the marine space, "drones" is not in common use, and "vehicle" is the most common equivalent. These systems are also often considered robotics systems. While they may not look like the characters from science fiction, these "robots" are at work on and under the water's surface. A varying degree of "autonomy" is also employed. Some systems are fully controlled by human operators, like the aerial and land robots known from military applications, while others roam more freely. The nature of the many different types of UMVs will be explored further.

But first it is important to understand the technical challenges that make UMVs a different breed from their land and aerial cousins. The role of water, naturally, is the key environmental difference. Corrosion from seawater is easy to grasp and is a mechanical engineering challenge widely addressed by all unmanned systems designers. But the pressure of seawater is a more significant obstacle. The pressure of the ambient seawater increases one atmosphere for each ten meters of depth. This becomes apparent in the ears of a swimmer trying to free-dive any distance below the surface. Scuba divers are trained to manage the pressure and carefully track their body's exposure to pressure so as to avoid dangerous health impacts, especially the bends, a potentially fatal outcome of rapid depressurization of the blood. UMV designers must also take significant steps to ensure the functionality of any system deployed beneath the waves. The engineering approaches are diverse and often the focus of entire courses in universities. Here it is simply important to note that moving parts, connectors, and even simple seals to keep out the water all must be highly engineered and carefully maintained once deployed.

With water, fresh or salt, safely outside the UMV there remain two key challenges to the systems: communications and positioning. Today the average mobile phone user takes for granted wireless telemetry and satellite-derived positioning. Unfortunately, for the UMV designer those well-proved technologies fail to deliver under water. The utility of radio frequency (RF) and electromagnetic (EM) technologies under water varies greatly with frequency and water conditions. Again, whole university courses address this topic, but essentially the energy of such transmissions are absorbed over just one wavelength, reducing useful propagation ranges to under 100 meters. What the UMV community understands well is that the wi-fi, Global Positioning System (GPS) and Bluetooth tools of such utility to unmanned systems operating in air are unavailable in the UMV ecosystem. A host of technologies, mostly based on the propagation of acoustic sound waves in water, are used instead. The impacts of this are seen in the designs and operational concepts of modern UMVs, which must use cables to provide the equivalent connectivity enjoyed by satellite-linked "drones." Hence the field of UMVs began with cabled systems.

7.2 Remotely Operated Vehicles

The oldest of the UMV family, remotely operated vehicles (ROVs), have assisted in ocean operations for decades. The basic concept of an ROV is that it uses a physical connection to bring power and control signals from the vehicle to the operator. Initially, these "tethers" were limited by the physics of copper conductors. Very long cables, to reach the deep ocean, demanded high voltages and offered limited bandwidth for signals and data, such as video feeds, the most common need of undersea operations being to see beneath the waves. As fiber-optics became available, tethers could support ever higher data rates, allowing for the introduction of advanced sensors and payloads on the ROVs. But managing the impacts of the tether remains a key aspect of ROV design and operation.

Early ROVs were often called eyeballs on a string: they used simple thrusters to move a few cameras and lights through the water. This approach carries on in systems commonly called inspection class ROVs. These systems are often quite small, in some cases able to be deployed by one individual hand-over-hand into the water. Despite their small size, these systems can support sophisticated instruments for hull inspection, seafloor imaging, and more. Police departments use small ROVs to locate drowning victims and others are used for scientific research on coral reefs. These compact members of the UMV ecosystem can be found nearly anywhere there is water. Usually, the most affordable of the ROV family inspection class ROVs can be obtained for as little as a few thousand dollars, and there are likely hundreds of manufacturers across the globe.

Less common, but substantially more capable, are work-class ROVs (Figure 7.1). These vehicles are designed for deep, hard work. They are often larger than a car and bristle with lights, cameras, manipulator arms, and tool skids. With thick, armored cables attached to tether management systems, or top hats, they are the workhorse of offshore industry. While these tools are immensely capable, they present one key operational challenge: they require vessels or platforms and crews. Some ROVs require a team of nearly 12 to sustain 24-hour operations. With a dynamically positioned support vessel in the mix, they become very expensive. But like airborne unmanned systems, all ROVs provide real-time connectivity to their operators. Pilots and the scientists or engineers working with them are able to inspect and manipulate the marine environment at a moment's notice. Most work-class ROVs are used in the construction and maintenance of subsea industry, such as oil rigs and pipelines. A smaller number are used by science for exploration and research and another handful are used by navies to support salvage and recovery of lost assets.

All ROVs can benefit from acoustic positioning systems. Two common types are ultra-short baseline (USBL), where a topside system measures range and bearing to a transponder on the ROV, or long baseline (LBL), in which seafloor transponders are used to triangulate the position of the ROV. Because the ROVs benefit from a hard-wired connection to operators, they do not require any wireless telemetry systems. They also can support sensors that produce significant amounts of data, such as multibeam sonars, which can provide 3D topographical maps of the seafloor.

Figure 7.1 A work class ROV. (courtesy of FMC Technologies—Schilling Robotics)

Regardless of size or class, ROVs are widely used by scientists, government agencies, and industry for a variety of missions, including pipeline inspection and maintenance, shipwreck exploration and recovery, hull and structure inspection, mine countermeasures, marine research, and undersea construction. ROVs are the undersea vehicles most like land and aerial systems owing to their real-time connectivity to human operators, but they are just the beginning of the UMV universe.

7.3 Autonomous Underwater Vehicles

While ROVs rely on a cable for power and data, autonomous underwater vehicles (AUVs) break free entirely. Also known as unmanned undersea vehicles (UUVs), AUVs carry their own batteries to support their operation. All necessary sensors and "intelligence" are built into the AUV.

A modern AUV will include the following elements:

- Energy storage: Often rechargeable batteries but sometimes fuel cells or primary batteries.
- Computing systems: A computer system collects inputs from the vehicle sensors, evaluates progress against stored mission programming, and issues commands to the vehicle systems.
- Propulsion: Usually a single propeller, sometimes multiple thrusters, and in some cases simply bladders to change buoyancy.

- Actuators: Often control fins near the thruster, though thrusters aimed across axes are also used.
- Sensors: Almost any undersea sensor developed has been tested on AUVs, commonly used systems are physical oceanographic instruments and mapping sonars.
- Positioning: Through a combination of onboard motion sensors, at least a compass, and external references, usually acoustic, the AUV can locate itself in the world.
- Telemetry: AUVs connect on deck and while floating using common wireless systems, but undersea they usually rely on acoustic modems to provide basic information on their activities. Acoustic telemetry does not offer enough bandwidth for a pilot to "drive" an AUV so they also require intelligence.
- Intelligence: The algorithms that allow AUVs to execute their tasks, sometimes for days at a time, are the heart of the system. There is a vast literature available on AUV control.

One could consider torpedoes the earliest of AUVs, with sophisticated mechanical control systems to ensure that the "vehicle" completed its serious mission. But in recent decades that same shape, long tubular with fins and propellers, has become a much more capable actor in ocean applications. With the power of modern computers and advanced software, AUVs can now support a large variety of missions. They also offer a variety of sizes and shapes. Small AUVs like those in Figure 7.2a can be carried by hand and used even in small lakes and ponds. Large systems like that in Figure 7.2b still must be deployed from vessels, but once under way the AUV can be left alone, allowing the vessel to perform other tasks.

Figure 7.2 Common examples of large and small AUVs. (courtesy of OceanServer Technology, Inc. and Bluefin Robotics Corporation)

As can be seen in Figure 7.2a and 7.2b, common AUVs look much like torpedoes. Although there is no clear common moniker for these, they are often called survey class AUVs. This derives from one of their most common applications: routine "mow the lawn" surveys over the seafloor. Driving back and forth carrying imaging sonars, or other sophisticated instruments such as sub-bottom profilers, survey AUVs provided information for pipeline route surveys, installation of bridges, clearance of minefields, or scientific research. Ranging from large systems supported by dedicated vessels to small units carried by hand, survey AUVs are now widely employed in ocean science, defense, and commercial applications.

Some survey AUVs are used to explore the vertical reaches of the ocean, where their pattern might change from boxes or grids to "yo-yos" up and down through the water. But there is another type of AUV that excels at this function, known as a glider (Figure 7.3). By changing volume it is possible to make objects rise and fall in the water. Scuba divers learn to adjust their buoyancy by controlling the amount of air in their lungs. Using the same principles, an unmanned vehicle can be made to sink or float. Once vertical motion is under way, the addition of wings can convert some of the vertical motion to horizontal. Just as gliders soar through the air without engines, so too can AUV gliders cruise very efficiently through the ocean.

Figure 7.3 An undersea glider. (courtesy of Ben Allsup, Teledyne Webb Research)

The primary advantage of gliders is very long endurance. The first UMV to cross an ocean was a glider, which sailed from the United States to Europe in 2009. With the possibility of staying at sea for months, perhaps even years, gliders provide an exceptional method for observing the physics, biology, and chemistry of the ocean. Their more affordable price, typically in the low hundreds of thousands of dollars, as compared with survey AUVs that can run into the millions, also makes them appealing to scientists and users who require multiple systems to achieve their missions. Today scientists and navies are adopting gliders rapidly, making them the most common type of AUV in use.

Much less common is an AUV that looks and moves like an ROV. By combining onboard batteries, sensors, and software with the multiple thruster–equipped body of an ROV, it is possible to make a UMV that can hover and slowly move around the seafloor, or man-made structures such as pipelines. These systems are seen as likely contributors to the growing offshore energy industry. Perhaps by docking with undersea recharging stations, these systems may become the worker bees of undersea industry. Figure 7.4 shows one early working example of this class, called by its manufacturer an autonomous inspection vehicle (AIV).

Figure 7.4 An autonomous inspection vehicle. (courtesy of Subsea7)

7.4 Autonomous Surface Vehicles

While AUVs and ROVs have been developed to face the challenges of operations below the water, autonomous surface vehicles (ASVs) operate on the surface (Figure 7.5). In contrast to their undersea brethren, ASVs—which are also sometimes called unmanned surface vehicles (USVs)—can make use of all the technologies applied in land and air systems. Thus, these vehicles can apply satellite telemetry and GPS and reduce the reliance on onboard intelligence. Precise positioning and real-time communications make it easier for operators to employ ASVs. The ability to use internal combustion engines can provide extended endurance. But as any sailor knows, exposure to wind and waves requires exceptional engineering design and mechanical construction. Despite these challenges, ASVs are now being applied to a variety of missions. They offer significant potential in military situations where removing personnel from danger is a common objective. They also can be a useful additive element, or force multiplier, to the tedious task of hydrographic survey, which demands methodical back and forth runs over the seafloor, much like mowing a lawn. Although the potential of ASVs is clear, they are challenged, more so than undersea systems, by the rules of the road.

Figure 7.5 An autonomous surface vehicle. (courtesy of ASV Ltd.)

7.5 Rules of the Road

UMV technology is always evolving. As the land and air unmanned vehicle communities have discovered, the regulatory environment does not always empower technology. While UAVs struggle to be accepted by the Federal Aviation Administration, California and Nevada have legalized driverless cars. Opinions of these circumstances vary widely, but there can be no argument that the policy environment impacts both fields. In the maritime sector, the technology and policy communities have been slow to engage. Early policy analysis of AUVs, for example, looked at impacts on marine life, not freedom of navigation [7-1]. In 2005, the U.S. Navy inspired an effort to develop technical standards for UMVs. Through a consensus-based process administered by the global standards organization ASTM, through Committee F41, a series of initial standards were completed, though these focused on technical issues [7-2]. During this timeframe, technologists evaluated the potential for "smart" UMVs to interact with other vessels at sea. Generally, these efforts concluded that it would be feasible for UMVs to "share the road" and follow the "rules of the road" at sea [7-3].

Since these initial technical and policy steps, there has been a gradual increase in interest in the policy regime for UMVs. In 2009, ASTM F41 established a subcommittee on rules and regulations and began an effort to develop best practices and possibly standards for UMV markings and behaviors. To date, no draft standards have been brought for review by the consensus- based process. But the ASTM effort has informed the U.S. Navy, which has begun to work in cooperation with the U.S. Coast Guard (USCG) and industry on the development of a regulatory framework for UMVs. This will include an approach to the global rules on the prevention of collisions at sea (known as the COLREGS), administered by the International Maritime Organization (IMO). The USCG has indicated it intends to approach this issue at the international level with the IMO.

While these efforts are under way, most UMV operators currently apply their own perspectives to their at-sea efforts, and no significant legal actions have arisen yet. It is reasonable to expect this situation to change. The lack of official regulation on UMV operations is unlikely to stand. Presuming regulations, or even best practices, are forthcoming, it is in the interests of all sides to develop a rational policy framework. Maximum benefit from UMVs will most likely be obtained if future policies do not treat all UMVs alike. It should present requirements appropriate to UMV size and available energy, lest major economic benefits be lost. Any regulatory framework should recognize the limitations of subsea connectivity (low bandwidth, but reliable) available to most UMVs. As all mariners know, operations at sea always entail risk. Sensible regulation manages risk to optimize return. The immense returns of UMVs await such a regulatory environment but are being achieved in practice today.

7.6 UMVs in Practice: Increasing Return on Investment in Ocean Operations

The ocean environment has always challenged the strength of technology and the skills of sailors. Today seafarers and members of the maritime community face an additional challenge; constrained budgets but growing mission requirements. Research budgets are being cut despite new concerns from climate change to ocean acidification. Civilian ocean agencies must guard against tsunamis and monitor hurricanes, but with declining resources. Vessel operators face volatile fuel costs while offshore industry must plan for increased regulations, especially in new sectors such as ultra-deep water and arctic regions. Even the military community must plan their investments carefully as they fulfill traditional roles while engaging new missions such as anti-piracy. In these types of circumstances, ocean operators must do more with less. Fortunately, UMVs offer increased return on investment when compared with more traditional vessel dependent techniques.

UMVs have been under development for decades, evolving beyond research labs and becoming commercial realities. Fortunes have been made and technical milestones thought beyond reach are proving no match for the increasing capabilities of today's UMVs. Currently, there are technically and commercially viable offerings in the major classes of UMVs and in the scientific, commercial, and military markets. These tools are now in routine use, and the operational impacts of UMVs can be assessed.

A framework for this assessment includes four themes: survey, inspection, observation, and connectivity. Survey can be described as methodical collection of data, typically about the seafloor, in a defined region. Scientists might survey in search of new reefs, Governments survey to ensure clear channels and navies survey to find mines. Inspection typically means close investigation of particular features, often with cameras or high-resolution sensors. In a military context, inspection might be for ship maintenance or security, whereas commercial operators routinely inspect the integrity of offshore structures. Observation includes monitoring of an ocean region, often over large distances, to systematically collect time series history of that region. Oceanographers turn to observation to understand trends in fisheries or climate, and military operators might be interested in undersea sound propagation. These different applications present varying temporal and spatial scales, ranging from days and meters to thousands of kilometers and years. Finally, connectivity is a key element of UMV operations. All operators deploy UMVs for a purpose; awareness of progress toward that purpose is a key requirement. Timeliness of information is also important to survey, inspection, and observation activities, even if complete data sets must await a physical return of the UMV. But the wait is usually well worth it. In addition to the expected knowledge of the seafloor, AUV surveys have discovered shipwrecks, revealed the presence of unexploded ordnance, and found new scientific discoveries.

7.7 Today's UMVs: A Stand-Alone Solution

Current UMVs can significantly improve the economics of ocean operations. Today's UMVs are allowing ocean operators to mitigate the impacts of declining budgets. The core UMV themes all reveal positive impacts on the cost and complexity of marine operations.

7.7.1 Survey

The first survey task UMVs addressed was deep-water surveys, especially in the oil and gas industry. AUVs allowed deep surveys to be executed by a single vessel, whereas previously they required two. This was a significant savings, but the market demand proved modest, and a handful of large AUVs appear to have satisfied the market [7-4]. The growth in the large/deep AUV survey market seems to have slowed, but new underlying technologies such as synthetic aperture sonar (SAS) continue to reduce the cost per square mile and thus bring the technology within reach of operators with smaller budgets [7-5]. Good area coverage rates at high resolution will allow for further baseline mapping of the seafloor. As anyone with a personal navigation device knows, the world is much easier to approach with a good map.

Although the large/deep survey AUVs bring added value, they still require dedicated vessels and sizeable teams. In the late 1990s, small ASVs first demonstrated the ability to collect high-quality survey data without the demand for large support vessels [7-6]. Today, compact, low-logistics AUVs provide big vessel performance at a small vessel price. Using affordable small vessels, often vessels of opportunity, such as local charter boats, survey operators can now conduct seafloor surveys at depths up to 1000 meters and count upon data quality equal to the larger usually more expensive AUVs [7-7]. Modular architectures allow diverse payloads and rapid battery changes. Now one AUV can meet multiple missions and maximize operating time and return on investment. Scenes such as that in Figure 7.6 will become more common as operators adopt and deploy these rapidly improving UMVs.

Figure 7.6 A low-logistics UUV reduces operational costs, sometimes eliminating the vessel. (courtesy of Teledyne Gavia)

7.7.2 Inspection

Undersea inspection began with divers, gradually replaced by ROVs. The ROV industry has grown to significant size, with approximately 1000 large work class ROVs in global service [7-8]. These ROVs are effective, but they are not exactly affordable. They are necessary, but significant investments that demand sophisticated vessels and trained operators are also required. Offshore industrial inspection has benefited from ROVs in that new missions can be accomplished. But dramatic improvements in the economics of the application are still slightly over the horizon, awaiting fully proven autonomous inspection vehicles (AIVs).

Meanwhile, small ROVs have matured rapidly and are changing the economics of many missions. Available in many sizes, shapes, and price ranges, these ROVs enable rapid and effective inspections of the seafloor and structures. Though depth ratings are not typically as great as work class ROVs, the missions within reach of smaller systems are benefiting from new technology. New sensors support applications such as leak detection and pipeline imaging, and improved auto-pilots and subsea positioning reduce the burden on pilots. The significant increase in the number of micro-ROVs deployed by police departments, universities, and other fiscally constrained operators is a testament to the economics of this type of UMV. While the user community is growing, the gap between the capabilities of work class ROVs and the micro-ROVs is closing. Select low-logistics ROVs now offer the high thrust-to-weight ratio required to support long tethers in high currents. Flexible architectures allow these ROVs to be adapted to powerful sensors without the high drag "strap-on" approach of micro-ROVs. Thus, a low-logistics ROV can now support missions such as bridge piling inspections in low visibility and high-current environments. A previously slow, expensive, if not impossible, task is now routine and economical. Likewise inspection of ship hulls for security, ship husbandry, or law enforcement (Figure 7.7) is enabled by the new generation of high-performance low-logistics ROVs.

Figure 7.7 Small ROVs improve economics, and safety, of marine law enforcement operations. (courtesy of Teledyne Benthos)

7.7.3 Observing

Ocean observing has traditionally relied on vessel-based measurements and moorings, which require significant infrastructure investments. But today, extended-endurance UMVs and autonomous profiling floats offer broad temporal and spatial coverage at low costs. Using buoyancy, thermal, or wave energy approaches, these new UMVs can remain at sea for months and years rather than hours and days. These systems have demonstrated endurance to cover large distances or loiter in place for extended deployments. Traditionally, physical oceanography has been the focus of these platforms, due to available sensors and limited sensor energy budgets. But today new biogeochemical sensors that can measure important properties such as pH and in situ carbon dioxide are being demonstrated on long-endurance platforms [7-9].

The most significant example of the economic impact of UMVs on ocean observing is the U.S. Navy Littoral Battlespace Sensing-Glider (LBS-G) program. This program is acquiring and deploying 150 gliders, in its first major phase, to support navy operational observations. Though the low capital and operational costs of these tools are a clear benefit to the economics of ocean observation, there is another, more subtle, effect to note. Academic or government programs often share data collected by these platforms freely. With hundreds of gliders and thousands of floats deployed, there is a wide user base to advance new concepts and technologies. Together, the sharing of data and wide user community make ocean observing UMVs highly leveraged investments, with benefits to individuals who are not directly utilizing them. Commercial fishermen and private beachgoers may not be operating UMVs, but they are using the information to shape their decisions.

7.7.4 Connectivity

Wireless networks and satellite navigation have revolutionized life ashore, and acoustic systems bring telemetry and positioning to UMVs. Empowered by advances in digital signal processing, simple tones and pings have evolved to broadband schemes. These techniques offer reliable connections between many platforms and in some cases also provide positioning information. Wise application of acoustic systems increases productivity and reduces risk. RF and optical systems are also emerging for use subsea. These technologies promise high bandwidth but over short ranges. The predominant mode for subsea connectivity is point to point, though adaptive networks have been demonstrated. Technologies are mostly closed with operators dependent on single sources or forced to own potentially redundant hardware. Subsea connectivity is having an impact, but it still evolving and penetrating the market.

7.7.5 Today's UMVs: A Glimpse of the Potential

UMVs are indeed reshaping the economics of marine operations. But most operators are recognizing only incremental benefits. The potential of UMVs can be viewed through the lens of mobile phones. The current capability base for UMVs is similar to the personal electronics scene of the late 1990s. At that time, a well-outfitted early adopter would have a stand-alone GPS receiver, mobile phone, and personal digital assistant. A lucky few managed to connect that phone to the personal digital assistant and access a rudimentary World Wide

Web, in text and grayscale. Similarly, today's UMV operators must acquire specific tools for particular missions. Modular architectures and shrinking sensors are increasing utility, but today we are still awaiting the equivalent of the smartphone era in ocean operations.

7.8 The Future: Networked UMVs

Continuing the comparison to mobile phones, what might UMV operators look forward to in the future? Today, an average smartphone owner can take their device around the world and access phone calls, e-mail, web sites, navigation, and a host of specialized apps. This capability is built upon enabling developments in hardware, especially energy storage and compact processors, and accepted standards for connectivity. Such a future is coming to the undersea realm (Figure 7.8).

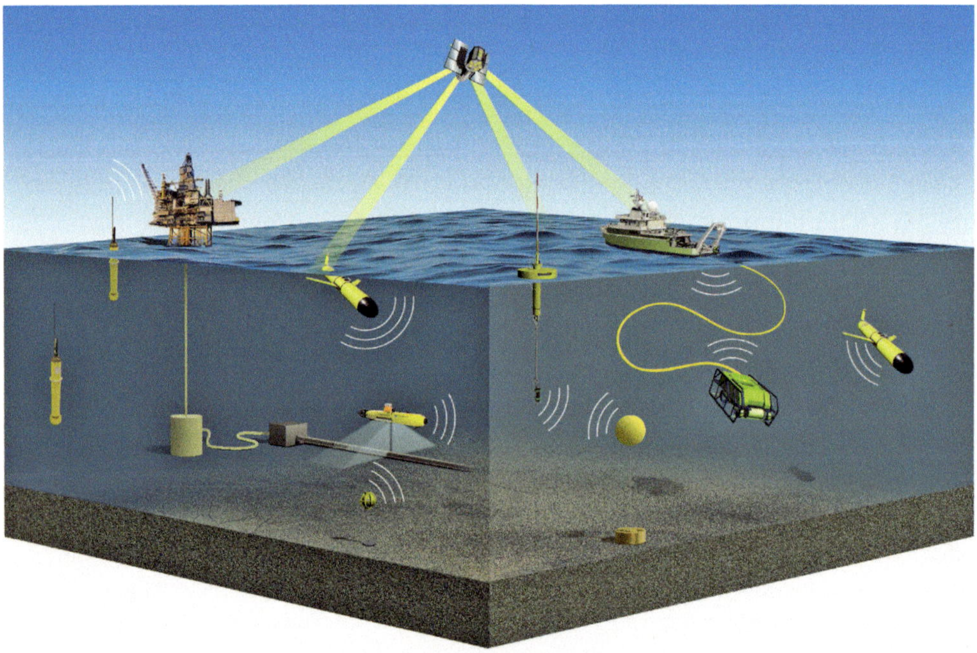

Figure 7.8 Networked undersea systems will yield new benefits to ocean operations. (courtesy of Teledyne Benthos)

UMVs in coming years will draw upon acoustic, RF, and optical connectivity to "network" into more productive roles. UMV payload modules will be interchangeable between a traditional survey AUV and a long-range glider, or a compact ROV. Meanwhile, gliders will offer propellers to overcome the occasional current or better navigate in operational areas such as under ice. AUVs will add buoyancy control to increase their efficiency or convert to drifting. Drifting and energy harvesting surface platforms will provide coverage for acoustic telemetry and positioning to heterogeneous undersea systems. How might such a future change the perspective for actual ocean applications? Consider the following hypothetical presentation of the four themes evaluated earlier.

7.8.1 Survey

An AUV is deployed from a rented fishing vessel, which simply launches it over the side. The AUV locates and follows a pipeline equipped with undersea "cell towers" that guide the AUV and provide operators ashore prompt notification of the AUV's findings. Pipeline condition is analyzed and anomalies flagged for follow-up without any dedicated vessel (significantly reducing cost). Meanwhile . . .

7.8.2 Inspection

A small, smart ROV is deployed from an affordable commercial support vessel. Sophisticated autopilots link the ROV and vessel, and the pair slowly follows the pipeline to inspect anomalies found by the AUV. A trained operator "points and clicks" the operation to completion, freeing highly skilled ROV pilots for intervention tasks (optimizing capital and operating investments). On large fixed installations, an autonomous inspection vehicle (AIV) deploys from a dedicated launch recovery and docking "basket" and provides wireless inspections of wellhead equipment. Meanwhile . . .

7.8.3 Observing

In open water, a drifting profiling float with advanced instruments detects a potential oil leak and adapts its dive to surface and alert operators ashore. They task a patrolling glider and nearby surface vessel (both energy harvesting and on station for months or years). The UMV pair home-in on the oil site and adaptively sample the plume, sharing the data with both commercial and government monitors. The regulators determine the event is a natural seep and response vessels are not required (saving both organizations valuable funds and time). Meanwhile . . .

7.8.4 Connectivity

Integrated networks of varied wireless technologies provide low bandwidth, but reliable coverage, much like first generation cellular networks. Full data sets will reside on local UMV solid-state drives, but system status and key data identified through onboard analytics will be exchanged across the network. Operators are informed and support assets are deployed for maximum efficiency. Consider the pipeline survey AUV, now docked with a node along the pipeline. To request recovery, the AUV can send the equivalent of an SMS request to predetermined regional assets, such as fishing vessels. The nearest available vessel can confirm the request on the network and move to recover the AUV. The recovery fee would more than compensate for the short deviation from fishing activity. Such an integrated subsea network minimizes downtime, even without the speed taken for granted in wired networks or atmospheric wireless systems.

7.8.5 Apps for Ocean Operations

Seafarers tend to be conservative and may find the vision presented preposterous. But those same seafarers today may video chat with their family from a vessel half a world away. In the not-too-distant future, after the video call on their smartphone, they could switch to a

UMV app and guide a sophisticated vehicle to its next mission; no cables, dive gear or joysticks required. The ocean is vast and will never be an easy operating environment, but with tomorrow's technology, today's tasks will be more like picking a restaurant on a smartphone than reeling in a net to a fishing dory.

7.9 Conclusion

Currently, UMVs are capable systems offering significant value to marine operations. The focus themes of survey, inspection, and observing provide a lens through which to view these tools and capabilities. The theme of connectivity, with its natural analogy to current mobile phone consumer experience, reveals that the potential value of UMVs is just beginning to be revealed. Just as smartphones reshaped daily lives, networked UMVs will bring more capability to marine operations. There are technical, business, and regulatory challenges to this vision. But none are insurmountable. The technology developments can be counted on to appear on scene, probably sooner than anyone expects.

7.10 References

7-1. Showalter, S. (2004). "The Legal Status of Autonomous Underwater Vehicles." *Marine Technology Society Journal,* Vol. 38, No. 1 Spring.

7-2. Lambert, J., and Manley, J. (December 2007). "Development of Unmanned Maritime Vehicle (UMV) Standards, an Evolving Trend," *Sea Technology,* Vol. 48, No. 12.

7-3. Showalter, S., and Manley, J. (October 2009). "Legal and Engineering Challenges to Widespread Adoption of Unmanned Maritime Vehicles." *Proceedings MTS/IEEE OCEANS 2009,* Biloxi, MS.

7-4. Kleiner, A., and Northcutt, J. (Jan/Feb 2004). "AUV Sea Trials and Tribulations–36 Months and 25,000 Kilometers Later, Was It Worth It?" *Hydro International.*

7-5. Wilby, A. (February 2012). "Deployment of a Deep Tow Synthetic Aperture Sonar System." *Sea Technology,* Vol. 53, No. 2.

7-6. Manley, J., and Vaneck, T. (November 1998). "High Fidelity Hydrographic Surveys Using an Autonomous Surface Craft." *Proceedings of Oceans Community Conference '98,* MTS, Baltimore, MD.

7-7. Hiller, T., Reed, T.B., and Steingrimsson, A. (November 2011). "Producing Chart Data from Interferometric Sonars on Small AUVs." International Hydrographic Review, No. 6, pp. 43-50.

7-8. Westwood, R., Westwood, J., and Newman, P. (2012). "The World ROV Market Report 2008 – 2012." Douglas Westwood Limited.

7-9. Fietzek, P., Kramer, S., and Esser, D. (September 2011). "Deployments of the HydroC™ (CO2/CH4) on Stationary and Mobile Platforms—Merging Trends in the Field of Platform and Sensor Development." *Proceedings MTS/IEEE OCEANS 2011,* Kona, HI.

Chapter 8
Unmanned Aircraft Systems in Environmental Monitoring Applications

Everett Hinkley, USDA Forest Service
Vincent Ambrosia, California State University-Monterey Bay / NASA Ames Research Center
Steven Wegener, Bay Area Environmental Research Institute (BAERI), NASA Ames Research Center

> Aviation in itself is not inherently dangerous. But to an even greater degree than the sea, it is terribly unforgiving of any carelessness, incapacity or neglect.
>
> Captain A. G. Lamplugh

8.1 Chapter Summary

Unmanned aircraft system (UAS) technology has gone through a significant evolution in the past 15 years. The current technology makes available significant capabilities that offer new ways of gathering information that in the past was impossible or difficult because of cost or safety considerations. The technology has advanced markedly over this period, and the necessary regulations that will allow these machines to operate in our national airspace (NAS) continue to mature. This chapter presents a taxonomy of unmanned aircraft systems and their application to a variety of environmental applications and will identify some pending engineering challenges and look ahead to what we might expect in the years to follow.

8.2 Introduction

Remote sensing gaps frequently exist over remote, scarcely populated and occasionally dangerous lands managed by the Forest Service, the Bureau of Land Management, and other civilian agencies. Resolving these data gaps through the use of either manned aircraft flights or satellite observations is often problematic. Manned aircraft flights are limited by crew duty cycles, are subject to unpredictable weather, and may have significant operational costs. Satellite-based observations are hindered by static sensor capabilities, weather conditions, and acquisition revisit cycles that are often measured in days or weeks. The Forest Service and NASA Ames Research Center have been working collaboratively to determine the effectiveness of UAS technologies in filling known observation gaps in land surface monitoring operations, with an initial focus on wildfire applications.

The development of UAS (also referred to as unmanned aerial vehicles [UAVs]) has been driven largely by military applications [8-1]. Though recent media coverage suggests that UAS are a recent innovation, unmanned flight has been around for many decades. Military applications have been a significant driver in UAS development; however, there are a host of nonmilitary applications to which UAS can be appropriately applied. This chapter will focus on resource monitoring and environmental applications for UAS technology.

Improvements in technology in recent years have increased interest among the broader scientific community in utilizing UAS for science missions in which unmanned aircraft provide clear benefits to risk mitigation. The National Aeronautics and Space Administration (NASA) pursued early advances in this technology and developed unmanned aircraft for atmospheric sampling during the 1970s and 1980s. The NASA Environmental Research Aircraft and Sensor Technology (ERAST) program in the 1990s took the first steps in developing an understanding of the protocols and capabilities needed for employment of UAS supporting scientific and other civil applications. The overarching goal of the ERAST program was to provide focus for critical technology development and flight demonstration that reduce the technical and economic risk of using UAS as a means to collect scientific data in a timely and cost-effective manner. At that time, UAS platforms capable of carrying desired remote sensing sensors tended to be large, risky, and expensive. Small UAS platforms lacked the ability to carry payloads with sufficient remote sensing capability to deliver desired data. One of the findings delivered by the ERAST program was the need for the development of smaller, more capable sensors to enable the utility of UAS platforms [8-1].

The late 1990s saw a number of efforts to develop UAS capabilities tailored to specific research needs, including atmospheric sampling and vegetation monitoring. Researchers quickly discovered that low-cost solutions were insufficient for many applications. In the early 2000s, the Forest Service began to evaluate UAS for active wildland fire monitoring applications. The Forest Service remote sensing experts at the Remote Sensing Applications

Center in Salt Lake City, Utah, teamed up with UAS experts at the NASA Ames Research Center in Mountain View, California, to leverage NASA's expertise in UAS and Earth science.

The understanding that UAS offer tremendous potential for many areas of scientific research and environmental monitoring has evolved rapidly over the past 15 years. This understanding has spurred significant demand in the public sector for systems that combine the sensor or sampling quality typically found on manned aircraft with the portability and cost advantages provided by unmanned systems [8-1].

As active members in UAS research and development, our intention is to introduce this technology to a readership who may be considering UAS as an alternative tool for remote sensing data collection and sampling.

8.3 Classification of UAS Platforms

Classification of UAS platforms for civil scientific uses has generally followed existing military descriptions of the platforms based upon characteristics such as size, flight endurance, and capabilities. The generally accepted class nomenclature in the civilian realm is as follows:

- **MAV (Micro [or Miniature] or NAV [Nano] Air Vehicles): These** are so called because of their size, which typically enables military versions of these aircraft to be transported within individual soldiers' backpacks. These aircraft tend to operate at very low altitudes (< 330 m), with size limitations on battery capacity leading to short flight times in the vicinity of 5–30 min.
- **VTOL (Vertical Takeoff and Landing):** These aircraft require no takeoff or landing run and are therefore typically chosen in situations where limitations of terrain require this specialized capability. Aircraft of this type operate at varying altitudes depending on their mission profile, but predominantly fly at low altitudes. High-power requirements for hovering flight limit the flight durations for VTOLs, except in the largest sizes, where increased lifting capabilities accommodate large fuel capacity.
- **LASE (Low Altitude, Short-Endurance): These** systems, also known as sUAS, small unmanned aircraft systems, also obviate the need for runways with aircraft optimized for easy field deployment/recovery and transport. The aircraft component of these systems typically weighs 2–5 kg, with wingspans < 3 m to enable launching from miniature catapult systems or by hand (Figure 8.1). Compromises between weight and capability tend to reduce endurance and communication ranges to 1–2 h and within a few kilometers of ground stations.

Figure 8.1 Cyberbug–LASE UAS. (courtesy of USDA Forest Service)

- **LASE Close:** This category describes small UAS whose aircraft do require runways, but whose larger size and weight confer increased capabilities. These systems operate at up to 1,500 m altitude and may remain aloft for multiple hours, although these limits have been substantially exceeded by specially modified "record-breaker" aircraft.
- **LALE (Low Altitude, Long Endurance):** Typically at the upper end of the "sUAS" weight designation by the U.S. Federal Aviation Administration (FAA), these UAS may carry payloads of several kilograms at altitudes of a few thousand meters for extended periods.
- **MALE (Medium Altitude, Long Endurance):** These aircraft are typically much larger than low-altitude classes of UAVs, operating at altitudes up to 9,000 m on flights hundreds of kilometers from their ground stations, lasting many hours (Figure 8.2).
- **HALE (High Altitude, Long Endurance):** These are the largest and most complex of the UAS, with aircraft larger than many general aviation manned aircraft. These UAVs may fly at altitudes of 20,000 m or more on missions that extend thousands of kilometers. Some HALE aircraft have flight durations over 30 h and have set records for altitude and flight duration.

A broad and rapidly evolving range of platforms together with sophisticated sensor packages are available to potential UAS users today. Those interested users who lack recent experience in this arena should expect a steep learning curve.

Figure 8.2 NASA Ikhana-MALE UAS. (courtesy of NASA)

8.4 Flight Regulations

It is important to note that an increasing number of diverse regulations affect the use of UAS in most countries, and these can be both confusing and discouraging to potential users.

8.4.1 Current Considerations in the United States

In the United States, there are two acceptable means of operating UAS in the NAS outside of "restricted" airspace: a Special Airworthiness Certificate–Experimental Category or a Certificate of Waiver or Authorization (COA). A Special Airworthiness Certificate in the Experimental Category is the only certification available to civil operators of UAS. Due to regulatory requirements, this approval precludes carrying persons or property for compensation or hire but does allow operations for research and development, market surveys, and crew training.

The COA process is available to public entities, such as government agencies (including local law enforcement and state universities), who want to fly a UAS in civil airspace. Applicants apply online, and the FAA evaluates the request [8-2].

A COA is required for flight in general airspace, and a Special Airworthiness Certificate (SAC) is required for flight in Military Operations Areas (MOA). Only Restricted and Warning Areas can be used without FAA authorization. Prohibited Areas also require specific authorization, since the intent of the Prohibited Area is for no aviation activity. All UAS operations in the United States require specific authorization (outside Restricted and Warning Areas). First responders must also meet this requirement [8-3].

Since 2006, the FAA has issued COAs to more than 100 unique public entities throughout the United States, ranging from local fire departments to research laboratories. Since the COAs have narrow scopes and durations, entities must obtain new authorizations every few months. In 2009, the FAA issued 146 such COAs. In 2010, the number of COAs issued grew to 298 [8-4].

8.4.2 Looking Ahead to Future Regulations

In February 2012, President Obama signed House Resolution 658, The FAA Modernization and Reform Act of 2012, into law, including Title III, Subpart B of H.R. 658 with language explicitly relating to future use and development of unmanned aircraft systems policy and guidance [8-5], [8-6]. This act sets the groundwork for integrating UAS into the U.S. NAS. It calls for development of a comprehensive plan to integrate all classes UAS in the NAS, in 270 days, including the following:

1. Final rules for the operations of smaller UAS (> 55 pounds) in the NAS
2. Notice of proposed rulemaking for the integration of all unmanned aircraft systems into the NAS system, with final rules in place by September 2015
3. Establishment of UAS test ranges to facilitate developing, testing, and validating of new technologies and procedures, including
    - Six national ranges coordinating mixed-use testing of UAS and manned aircraft
    - Development of certification standards and air traffic requirements supporting public and civil UAS
    - Ensuring coordination with the Next Generation Air Transportation system
    - Providing for the testing and validation of UAS and procedures

The act states: "The plan shall provide for the safe integration of civil unmanned aircraft systems into the national airspace system as soon as practicable, but not later than September 30, 2015" [8-2].

8.5 UAS Applications in Environmental Monitoring

The utility of UAS and related technologies is currently being evaluated by several civil agencies for monitoring and characterizing conditions of natural resources to support a variety of operational applications and information needs. The potential benefits of these technologies to the Forest Service include rapid and focused deployment, unique data acquisition and communications capabilities, analysis/display capabilities (sometimes in near real time), and reduced risks to flight personnel. The U.S. Geological Survey (USGS) and the Department of the Interior have acknowledged increasing value in the applications of UAS for environmental monitoring and evaluation of fast-changing hazards, such as wildfires and volcanoes, and monitoring ephemeral wildlife habitats.

UAS and associated technologies are rapidly emerging and hold potential to support operational needs across several organizational areas in the Forest Service. Consequently, the ongoing rapid advancement in UAS and associated technologies, and the requirements for UAS operations to adhere to aviation regulations, necessitate an agencywide approach to coordinating and managing the operational implementation of UAS in the Forest Service. In order to promote appropriate development and use within the agency, the Forest Service has established the Unmanned Aircraft Systems Committee to ensure that Forest Service UAS planning and operations enhance the ability of the agency to perform its mission more efficiently and effectively.

8.5.1 Wildfire Applications

Over the past decade, the Forest Service has conducted a wide range of demonstration projects to assess the suitability of UAS for support of wildland fire management. Data collection in and around a fire incident is an inherently dangerous and expensive business, and these circumstances provide an attractive niche for the application of unmanned aircraft (Figure 8.3).

Figure 8.3 UAS over wildfires concept. (courtesy of NASA)

There are a host of benefits to the use of UAS in wildfire applications:

- Decreases hazards to support staff by removing them from dangerous situations.
- Provides synoptic landscape viewpoint and repeat coverage compared with ground observations, especially in areas that are difficult to access with manned aircraft.
- Provides the ability to observe for extended periods over areas of interest to support day/ night operations beyond crew duty cycles.
- Projected lower costs to operate (small UAS) compared with manned aircraft. Though this has yet to be demonstrated with UAS, all technologies have a development curve, and economics of scale will come into play in the future of UAS technology. Small owned UAS systems offer anticipated lower costs than manned aircraft, though there is a trade-off as image quality suffers in small payload capable sensors.
- Puts desired observational capabilities in the hands of the incident command units to fly when and where needed.
- Agile observational capabilities provide additional opportunities for situational awareness to aid in informed decisions.
- Real-time monitoring from a distance.

These capabilities were readily demonstrated during missions organized by the USFS and NASA from 2006 to 2010 involving the NASA Ikhana UAS platform (Predator B derivative). This mission series demonstrated safe, efficient operations and real-time data relay from an onboard sensor system through satellite telemetry systems to incident management teams on the ground. The successful mission series, operated over eight western states, collected data on over 60 fires, during a total of over 180 hours of operations [8-7].

8.5.2 Forest Health Monitoring

Our forests are environmental assets that belong to all of us, and the world's forests are under siege from climate change and severe pest and disease infestations. It is important that we monitor and report changes of the health of the forests. The Forest Service currently uses manned aerial surveys to track the health of our forested lands on state, federal, and private lands (Figure 8.4). This is a challenging and often dangerous activity, and the Forest Service is interested in conducting future UAS missions to assess forest survival conditions and to determine stocking levels in pest/disease-impacted areas located in remote and/or unsafe areas. The use of UAS in forest health monitoring requires advanced electro-optical (E/O) sensor payloads, such as near-infrared and, in some cases, hyperspectral sensors.

Figure 8.4 Manned flights over forested areas. (courtesy of USDA Forest Service)

We envision a time when forest surveys around the world may be done safely and effectively with UAS. For example, the United Nations may want to consider commissioning a fleet of UAS to inventory and track the health of the planet's forests. Due to extensive areas of coverage required, MALE UAS are likely to be the desired platform for these missions.

8.5.3 Wildlife Surveys

UAS-enabled observations of wildlife have been conducted primarily using LASE UAS (Figure 8.5), whereas MALE aircraft may be preferred for marine surveys due to requirements for extended operations. In 2009 in Alaska, the University of Alaska–Fairbanks (UAF) employed a Boeing-In Situ ScanEagle UAS to identify ice seals along coastal regions of Northern Alaska. Those same platforms have also been used by UAF to survey whale populations in Alaska coastal waters to study their environments and migratory patterns.

The National Oceanic and Atmospheric Administration (NOAA) has also completed assessment missions with UAS in support of test-bedding capabilities for marine mammal inventories utilizing a MALE aircraft, a modified Predator-B (Altair platform). Those mammal inventory and marine enforcement surveillance missions were performed in April–November of 2005 over the Channel Islands National marine Sanctuary [8-8].

The Department of Interior, in conjunction with the USGS National Unmanned Aircraft Systems Project Office, routinely uses sUAS to assist DOI agency personnel in wildlife monitoring capabilities. Some recent examples include Sandhill Crane population estimates in Colorado and Pygmy Rabbit habitat mapping in Idaho. Planned missions include Seagull Monitoring in California off-shore habitats (Farallon Islands) and monitoring wildlife and habitats on Palmyra Atoll in the North Pacific Ocean [8-9].

Figure 8.5 The Polaris/Nova UAS is launched from an airboat in Florida's Everglades. (courtesy of Adam Watts)

Each of these missions was established to test, evaluate, and demonstrate new capabilities of UAS to support wildlife management objectives and to form the foundation for implementing operational capabilities into wildlife management survey techniques.

8.5.4 Air Quality Monitoring

Sampling payloads may require the additional lift capacity of larger UAS, but generally the geographic scope of the mission profile will determine whether short-duration flights by VTOL or the greater coverage and duration offered by LASE or MALE UAS are preferred.

In November 2005, the U.S. National Oceanic and Atmospheric Administration (NOAA) completed its last in a series of missions over the eastern Pacific Ocean with the unmanned Altair aircraft. The flight was part of its Unmanned Aircraft System (UAS) Demonstration Project and initiated the era of unmanned flights in service of environmental missions. Altair cruised at altitudes in the lower stratosphere (13 kilometers; ~43,000 feet), collecting atmospheric data with a 140-kg payload of both remote sensing and in situ sampling instrument systems.

NOAA recognized that UAS technology would improve its ability to meet scientific and operational objectives. Operating sensor payloads on a UAS fleet could play a crucial role in the detection and attribution of climate change, improvement of weather predictions, management of water resources, monitoring and evaluation of ecosystems and sanctuaries, and atmospheric and oceanic research [8-10].

8.5.5 Agriculture and Crop Monitoring

Two critical limitations for using current satellite sensors in real-time crop management are the lack of imagery with optimum spatial and spectral resolutions and an unfavorable revisit time for most crop stress-detection applications. Alternatives based on manned airborne platforms are lacking due to their high operational costs. A fundamental requirement for providing useful remote sensing products in agriculture is the capacity to combine high spatial resolution and quick turnaround times. Remote sensing sensors placed on UAVs could fill this gap, providing low-cost approaches to meet the critical requirements of spatial, spectral, and temporal resolutions [8-11].

Monitoring of crop health and precision application of chemicals require large payloads and advanced E/O sensors, as well as precise navigational and maneuvering capability. Relatively small areas of operation combine to yield a preference for larger VTOL UAS in these applications.

UAVs and other small unmanned aircraft have potentially lower cost but also have a significantly lower payload capacity, so light-weight compact sensors are required. Furthermore, UAVs can be flown at lower altitudes with greater safety than manned aircraft, thereby increasing spatial resolution. Low-cost, light-weight sensors are critical for the development of UAVs as a cost-effective platform for image acquisition [8-12].

8.5.6 Water Science

For monitoring of water pollution, repeated flights from local platforms may be accomplished by VTOL UAS. However, extensive surveys such as those required to track oil spills may make MALE UAS preferable. Payload requirements, which will impact platform choice, will be driven by specific pollutant types.

Remote sensing requirements for monitoring lakes, rivers, and ocean systems and advancing water science present a number of opportunities for UAS:

- Global climate change and ice cap monitoring (Figure 8.6).
- Monitoring CO sinks in the ocean.
- Looking at ocean near surface (10 m above the ocean) exchange rates.
- Kelp bed monitoring, algal blooms, and analysis of red tide events.
- Monitoring oil spills: During the 2010 Deepwater Horizon oil spill event in the Gulf of Mexico, the Department of Homeland Security (DHS), Customs and Border Protection (CBP) provided Predator UAS platforms for mapping the oil extent using onboard imaging sensors. Although the platforms were removed from the monitoring effort after two-weeks due to their expense in operations, it exemplified the use of a national asset UAS platform to map extensive areas affected by a disaster event [8-13].
- Fishery monitoring.
- Flooding.
- Littoral zone monitoring.

Figure 8.6 The NASA SIERRA, depicted flying near Svalbard, Norway, 2009. (courtesy of NASA)

8.5.7 Disaster Event Monitoring

Emergency response in large disaster scenarios presents a number of very compelling opportunities for UAS, both in active-event monitoring and in post-event assessment.

- **Hurricane monitoring:** NASA and NOAA have collaborated to monitor hurricane genesis and movement in the Atlantic utilizing the Global Hawk UAS platform. The purpose of those missions is to observe the genesis of the hurricanes near the Cape Verde Islands in the eastern Atlantic Ocean, off the African coast. The platforms are capable of tracking/following the hurricanes through their early-stage developments and enable an increased understanding of the hurricane path movements and changes in wind speed/direction, including the influences of water currents and other climactic factors. With the long-duration capabilities of HALE UAS like Global Hawk, a long-term (~32-hours) monitoring capability is enabled.
- **Temporary communication links**: Because of the potential for higher altitude and long-duration operations, UAS are being considered to fill the role of a communications system or communications "relay-in-the-sky" during events that damage or incapacitate ground-based communications networks (cell towers, infrastructure, and so on) An example of this was the SkyTower Telecom System Concept developed by AeroVironment utilizing the Global Observer UAS platform. The concept was to provide broad-area coverage of two-way broadband, broadcast video/audio, and narrowband voice communications over large affected areas (50–250-mile diameter) from a platform operating at 65,000 ft, above weather and air traffic, for over one-week flight duration [8-14].

Small UAS (sUAS) can also play a role as short-term communications relay platforms during disaster response events. During situations where emergency management teams need to stay in communications contact in "communications-blind" areas, an sUAS can serve the needs as a radio-repeater platform. An example of this use was demonstrated by the USFS and NASA in 2006, when an APV-3 sUAS platform (Figure 8.7) was fitted with a USFS radio repeater payload and served as a communications relay between a wildfire incident command team and a "forward" firefighting unit. The firefighting unit was located more than 5 miles from the command center in a terrain-induced radio "shadow" area. This situation necessitated an elevated repeater station to allow the forward unit to communicate with the command center. The system operated very well in these conditions.

Figure 8.7 RNR APV-3 with belly pod carrying communications relay package, 2006. (courtesy of USDA Forest Service)

- **Search and rescue (SAR)**: There is growing interest in sUAS utility in providing quick SAR support. This is predicated on experience gained in military applications of sUAS for such missions. A hand-launched sUAS can be quickly deployed in extreme terrain/ground cover environments to serve as an "eye-in-the-sky" and look for lost personnel (hikers, climbers) by way of both real-time thermal video feeds (identifying "warm" personnel in cool backgrounds) or standard color video, where the SAR victim is spectrally differentiated from background vegetation. These quick-launch capabilities allow rapid assessment of urban environments as well, allowing focused attention on specific areas where SAR operation might initiate, such as during floods, tsunamis, and earthquake events.
- **Determination of extent, magnitude, and triage of infrastructure damage**: The stand-off capabilities of UAS platforms serve them well for making rapid determinations of extent and magnitudes of damage from disaster events. Two opposite-spectrum cases serve as excellent examples of this capability. In 2011, following the devastating earthquake and accompanying tsunami in Japan, both small, tactical UAS and large,

strategic UAS were employed for hazard assessment at the Fukushima Daiichi Nuclear Power Plant. The U.S. Air Force deployed a Guam-based Global Hawk to assess damage to the plants. Additionally, small UAS were used at Dai-Ichi as well to make close-up inspections of damage. These platforms included a Honeywell Aerospace RQ-16B T-Hawk UAS, a Fuji IMVAC company B-Type Machine II platform, and others. These real-time video, close-up assessments allowed fine-resolution imagery to be taken, allowing a "stand-off" capability in a life-threatening environment.

- **Real-Time Input for Decision Support Systems (DSS):** Real-time information from UAS platforms allows integration/use in geospatial analysis systems/GIS systems and therefore reduces the time for intervention on the event. Derived information can be integrated seamlessly into the disaster management infrastructure, decision support system, to save resources critical time and allow an improved/faster action plan to be developed. The data derived from the UAS sensor systems are now being distributed in Open Geospatial Consortium (OGC)–compliant formats, allowing a simple ingestion into most custom/public-developed DSS systems.

8.6 Engineering Challenges

8.6.1 Safety – "Sense and Avoid"

To fly UAS in shared airspace with conventional manned aircraft is inherently dangerous when the UAS lacks the ability to "sense" and react to other aircraft. The understanding of other aircraft presence, location, and heading is necessary to respond to developing conflicts. The capability to react is termed "sense and avoid" (or "see and avoid") and is a key component of a UAS-enabled future.

There are a couple of ways to accomplish sense and avoid. The project team can implement a ground-based observational capability to communicate potential conflicts to the pilot in order to initiate evasive actions (FAA National Airspace System, or ground observers for close-in applications). Additionally, the user may employ other technologies (e.g., radar, transponders, and perhaps other technologies) to detect and track other aircraft in order to assess risk and to trigger evasive maneuvers (UAS autonomous or by ground pilot). New sense-and-avoid technologies are an area of intense research and development at this time.

8.6.2 Communications

The Federal Communications Commission regulates the nation's radio frequency environment. Access to unused radio frequency (RF) spectrum is highly competitive across several industries. The RF spectrum is particularly contested in and around urban centers, with congestion scaling upward with population. As the operational area and/or altitude of UAS expands, so scales the requirement for communications planning and expanded spectrum management.

Communications are a critical element of UAS operations. UAS depend on RF communications for command and control, systems monitoring, and payload operations. Although most UAS have autopilots that are capable of managing scripted mission plans, most launch and recovery operations must be done by pilots controlling the aircraft via low-latency line of site radio systems. Modifications to autopilot mission plans, aircraft systems monitoring, and payload operations are also done via RF systems, applications that are more tolerant to latencies typical in networked communications, enabling autonomous operations, even beyond line of sight on commercial networks (i.e., cellular or satellite).

The communications requirements of payload are highly variable. Minimal requirements might provide for just sensor status, though some applications might call for continuous transmission of high-resolution imagery. Many users may require sensor data in near real time for urgent applications supporting streamlined decision support systems (law enforcement, infrastructure management, disaster management, news and traffic) that cannot wait for the data to be downloaded and analyzed after the aircraft lands.

Other considerations:

- For improved safety, all vehicles operating in the NAS need transponders transmitting identification and location information.
- The pilot-in-command needs to be in touch with local air traffic controllers, either directly or via a "beyond line of sight" network paired with onboard standard VHF radios.
- Some applications may desire to incorporate a communication-triggered flight termination system, requiring secure and reliable radio links.
- A clean and secure RF environment is critical for safe operations, requiring a well-thought-out communications architecture, based on managed spectrum best practice.

The pressure on the useable RF spectrum will continue to increase with the introduction of new technologies and an expanding user base.

8.7 The Next Ten Years

Much like the enabling benefits of the Internet and the Global Positioning System, UAS have demonstrated powerful capabilities that will enhance our ability to be better stewards of the land and seas. The past decade's rapid evolution of UAS platforms and sensors has yielded attractive capabilities to a diverse end user community looking for efficient and agile data acquisition along with reduced costs. The ongoing maturation of increasingly miniaturized and sophisticated sensor components is leading to an era of tailored, capable systems for on-demand remote sensing in environmental monitoring applications. While operational mission costs are difficult to accurately assess at this stage in their development, it is clear that these costs are coming down and will continue to do so as the technology gains a foothold as a reliable, operational remote sensing tool.

The capabilities found in UAS will transform natural resources monitoring and management in significant ways. UAS will serve a variety of data-gathering needs, including disaster monitoring, natural resource assessments, habitat assessments, climate studies, and a myriad of other uses. The utility of environmental applications has been clearly illustrated in a number of demonstrations on unmanned aircraft of varying size and capability. Data acquisition in areas previously inaccessible or of limited access is made easier and more efficient. Data capture rates are enhanced with more rapid acquisition cycles of information from quick-launch, and long dwell, on-site platforms.

As UAS operations for nonmilitary purposes become routine, this technology will continue to evolve and expand to an ever larger pool of potential uses and users. Government agencies or universities will likely remain among the largest users of UAS. Private vendors are expected to be the largest providers of UAS equipment and services. In 2011, the FAA reported that in the United States alone, approximately 50 companies, universities, and government organizations are developing and producing over 155 unmanned aircraft designs [8-2].

Commercial UAS manufacturers will increasingly tailor their products to the needs of civilian clients as wartime military outlays for UAS stabilize and peacetime demand increases with regulatory clarification and acceptance. Additionally, the service sector will evolve to offer UAS services such as leased systems, on-demand flights, or consultation for choosing appropriate platforms or analyzing UAS-generated data. There is little doubt that the civil UAS service sector will see dramatic growth in the coming decade. Additional availability of UAS through the acquisition of military surplus by approved users (civil government agencies) may provide some benefits through greater familiarity with UAS use and operation, although it can be argued that the scientific benefits of surplus UAS are diminished in the face of the rapidly evolving technology.

The potential user of UAS technology will need to take care in considering their data requirements, which will drive the decision on which sensors are best to carry out their mission. Payload size, weight, and power draw coupled with flight duration requirements and operational cost will largely determine the appropriate aircraft acceptable for the mission. Choosing unmanned versus manned aircraft is a decision that must also be made with care. With current cost per hour pricing, manned aircraft often offer the cheaper and more efficient alternative in many instances; however, cost is seldom the only consideration. For example, in fire and disaster response, pilot and crew safety must also be considered along with human flight limitations.

8.8 Conclusion

- While UAS continue their rapid technical evolution, regulatory bodies are diligently working to develop regulations and guidelines to safely manage this new technology while balancing the economic, information, and safety needs of the user community and the public. Once airspace guidelines and access issues are in place (by the FAA in the United States, for example), the technology will see rapid adoption in many environmental applications.
- The future of UAS in support of environmental monitoring is assured. We are witnesses to a new technology exploding on the scene that significantly augments our current capabilities. As this chapter goes to press, the NASA Global Hawk is surveying

Hurricane Leslie in the Atlantic, making long-duration measurements and observations from 65,000 feet above the ocean surface. With its 32-hour flight endurance, the Global Hawk allows the community unprecedented access to open-water tracking of hurricanes to increase our understanding of storm dynamics, rates of movement, wind patterns, and pressure systems and will allow an improved understanding of landfall areas and human impact. This was unachievable just three years ago, but we are living in a new era of flight, the era of UAS. And we can say with confidence: the future is here!

8.9 References

8-1. Watts, A.C., Ambrosia, V.G., and Hinkley, E.A. (2012). "Unmanned Aircraft Systems in Remote Sensing and Scientific Research: Classification and Considerations of Use." *Remote Sensing*, 1671-1692.

8-2. Dorr, L., and Duquette, A. (2010, December 1). *Fact Sheet–Unmanned Aircraft Systems (UAS)*. Retrieved August 23, 2012, from Federal Aviation Administration: http://www.faa.gov/news/fact_sheets/news_story.cfm?newsid=14153 Corrected/updated link, retrieved May 2, 2014.

8-3. Williams, A. (2012, July 2). E-mail correspondence. Washington, DC, USA.

8-4. Elias, B. (2012). *Pilotless Drones: Background and Considerations for Congress Regarding Unmanned Aircraft Operations in the National Airspace System*. Washington, D.C.: Congressional Research Service.

8-5. 112th Congress—House of Representatives. (2012, February 1). *FAA MODERNIZATION AND REFORM ACT OF 2012 Conference Report to accompany H.R. 658*. Retrieved August 31, 2012, from GPO.gov: http://www.gpo.gov/fdsys/pkg/CRPT-112hrpt381/pdf/CRPT-112hrpt381.pdf.

8-6. Government Printing Office. (2012, February). *House Resolution 658*. Retrieved August 23, 2012, from Government Printing Office: http://www.gpo.gov/fdsys/pkg/BILLS-112hr658enr/pdf/BILLS-112hr658enr.pdf.

8-7. Ambrosia, V.G., Wegener, S., Zajkowski, T., Sullivan, D.V., Buechel, S., Enomoto, F., et al. (2011). "The Ikhana UAS Western States Fire Imaging Missions: From Concept to Reality (2006-2010)." *Geocarto International Journal*, 85-101.

8-8. Aslaksen, M. (2006). *The NOAA Unmanned Aerial System (UAS) Demonstration Project Using the General Atomics Altair UAS*. National Oceanic and Atmospheric Administration.

8-9. USGS. (2012, August 17). *National Unmanned Aircraft Systems (UAS) Project Office—Mission Listings*. Retrieved August 30, 2012, from USGS—National Unmanned Aircraft Systems (UAS) Project Office: http://uas.usgs.gov/missions.shtml?current=3.

8-10. Fahey, D.W., Churnside, J.H., Elkins, J.W., Gasiewski, A.J., Rosenlof, K.H., Summers, S., et al. (2006). Altair Unmanned Aircraft System Achieves Demonstration Goals. *EOS Transactions*, 197-201.

8-11. Berni, J.A., Zarco-Tejada, P.J., Suárez, L., and Fereres, E. (2009). Thermal and Narrowband Multispectral Remote Sensing for Vegetation Monitoring From an Unmanned Aerial Vehicle. *IEEE Transactions on Geoscience and Remote Sensing.*

8-12. Hunt, E.R., Hively, D., Fujikawa, S., Linden, D.S., Daughtry, C.S., and McCarty, G.W. (2010). "Acquisition of NIR-Green-Blue Digital Photographs from Unmanned Aircraft for Crop Monitoring." *Remote Sensing*, 290-305.

8-13. Olive, D. (2011, March 22). *Homeland Security Policy Institute—Security Debrief.* Retrieved August 23, 2012, from CBP'S Predator UAV Use Raises More Questions, Answers Elusive: http://securitydebrief.com/2011/03/22/cbps-predator-uav-use-raises-more-questions-answers-elusive/.

8-14. *Global Observer.* (2012). Retrieved August 23, 2012, from Aerovironment: http://www.avinc.com/globalobserver.

Chapter 9
Privacy

Brett Davis, AUVSI

> Unmanned aircraft have the potential to help law enforcement agencies with missions such as search and rescue or crime scene photography, often at a lower cost than manned aircraft. While there certainly should be a reasonable conversation about the application of any new technology, we are concerned by any proposed legislation that might unnecessarily limit law enforcement agencies' ability to use unmanned aircraft to keep communities safe.
>
> Michael Toscano, president and CEO of AUVSI.

9.1 None of Your Business: Unmanned Systems and Privacy

When the U.S. Congress passed and the president signed into law legislation directing the Federal Aviation Administration to pave the way toward the integration of unmanned aircraft systems into the U.S. national airspace, it represented a major milestone for a burgeoning technology.

The bill was the Federal Aviation Reauthorization Modernization and Reform Act of 2012, and it was signed by President Obama on February 14, 2012. Passage of the legislation was not a rapid process. The bill included many components and overall required 23 short-term extensions. However, once approved, it made history by setting unmanned aircraft on track to fly in the National Airspace System of the United States.

Capable of performing dangerous and difficult tasks safely and efficiently, unmanned aircraft systems (UAS) could soon assist search and rescue teams trying to find a missing child. They could soon help firefighters battle wildfires by detecting hot spots and relaying images in real time. Or they could soon help scientists better understand the paths of hurricanes or the plumes of volcanoes and offer critical information to keep people safe from natural or man-made disasters.

Some small, unmanned aircraft were used as far back as 2005 in the wake of Hurricane Katrina, using their heat-sensing imagers to search for survivors in the wreckage. They have been used in many similar situations since. Unmanned aircraft were used by private industry and government alike in the wake of Haiti's devastating earthquake in 2010, including a high-flying U.S. military Global Hawk, which provided aerial imagery of the damage. Unmanned aircraft and ground robots were employed in the wake of the tsunami and subsequent Fukushima Daiichi nuclear plant disaster in Japan in 2011. There, radiation made it too dangerous for humans to get near, but small aircraft and ground robots were able to survey the damage.

Not only does the integration of UAS into the national airspace mean extending human potential in these and other ways, but it also could mean a significant financial impact on state and local economies across the United States. A 2013 report commissioned by the Association for Unmanned Vehicle Systems International (AUVSI) on the economic impact of the industry found that in the three years following the integration of UAS into the national airspace, the industry could help create 70,000 jobs and make $13.6 billion in economic impact. Beyond the first three years, the study projects that more than 100,000 new jobs will be created by 2025, as well as have $82.1 billion in economic impact [9-1]. Such economic impact could be duplicated around the world. In a study released at the 50th annual Paris Air Show, the Teal Group predicted that the worldwide market for unmanned aircraft alone will total slightly more than $89 billion over the next decade [9-2].

9.1.1 Addressing Concerns

But while the economic potential and societal benefits are clear, the prospect of the wider use of UAS has raised concerns as well. Like any new technology that is not commonly known or used, there are always questions about how that technology will be employed. In this case, as the expanded use of UAS was brought closer to reality by the FAA bill and by similar regulatory movement in other countries, issues on how UAS would be used and by whom have created a steady conversation in the media about this technology's impact on personal privacy.

Many of these concerns are perfectly reasonable. All people cherish their personal rights to privacy, and Americans value the rights guaranteed to them in the Fourth Amendment to the U.S. Constitution against unreasonable search and seizure. With the use of UAS set to increase dramatically in the years ahead, fueling apprehensions about privacy, AUVSI recognized the need to address the issue proactively and did so in several ways. Now in its 42nd year, AUVSI is the world's largest nonprofit devoted to advancing the unmanned systems and robotics community. In recent years, it has sought to educate the general public and lawmakers about the benefits of unmanned systems and robotics. The relatively sudden groundswell of unease about unmanned systems and privacy threatens to slow or even stop the societal and economic benefits expected from these systems, so the association felt it was imperative to act.

Numerous groups around the world have staked out positions regarding privacy, many wanting to strictly limit the use of unmanned systems—particularly aircraft—in addition to seeking limits on how the data is actually used. As the debate unfolded, AUVSI reached out to nearly a dozen privacy advocacy organizations and civil liberties groups, including the American Civil Liberties Union and the Center for Democracy & Technology, to hear their concerns.

9.1.2 Proactive Guidelines

In an attempt to deal with these aspects up front, AUVSI and the International Association of Chiefs of Police (IACP) separately issued guidelines on how unmanned aircraft should be operated, both of which dealt to some extent with the issue of privacy. AUVSI published its Unmanned Aircraft System Operations Code of Conduct in July 2011, a set of guidelines to provide AUVSI members—and those who design, test, and operate UAS for public and civil use—with recommendations for their safe, nonintrusive operation. This code is meant to provide UAS industry manufacturers and users with a convenient checklist for operations and a means to demonstrate their obligation to support the growth of the industry in a safe and responsible manner.

By proactively adhering to these guidelines, AUVSI reinforced how the rights of individuals and the safety of all users of civil airspace are a top priority. AUVSI acknowledges the importance of the issue in the preamble to the Code of Conduct [9-3]:

"The emergence of unmanned aircraft systems (UAS) as a resource for a wide variety of public and private applications quite possibly represents one of the most significant advancements to aviation, the scientific community, and public service since the beginning of flight. Rapid advancements in the technology have presented unique challenges and opportunities to the growing UAS industry and to those who support it. The nature of UAS and the environments which they operate, when not managed properly, can and will create issues that need to be addressed. The future of UAS will be linked to the responsible and safe use of these systems. Our industry has an obligation to conduct our operations in a safe manner that minimizes risk and instills confidence in our systems."

Central to the Code is the need for "safety, professionalism and respect" in all uses of UAS. The Code of Conduct states:

- *We will comply with all federal, state, and local laws, ordinances, covenants, and restrictions as they relate to UAS operations.*
- *We will respect the privacy of individuals.*
- *We will respect the concerns of the public as they relate to unmanned aircraft operations.*
- *We will support improving public awareness and education on the operation of UAS.*

In addition to the Code of Conduct, AUVSI released a "privacy statement" regarding the use of UAS [9-4]. The statement read, in part:

The Association for Unmanned Vehicle Systems International (AUVSI) supports the development and advancement of UAS technology in a safe and responsible manner, while respecting existing privacy laws and ensuring transparency and accountability. To help safeguard Americans' right to privacy, AUVSI supports:

- *The registration of unmanned aircraft and pilots with the Federal Aviation Administration.*
- *The enforcement of established law and policy, governing the collection, use, storage, sharing and deletion of data, regardless of how it is collected.*
    - *These policies should be available for public review.*
    - *The policies should outline strict accountability for unauthorized use.*
- *The International Association of Chiefs of Police recommended guidelines (http://www.theiacp.org/portals/0/pdfs/IACP_UAGuidelines.pdf) for UAS operations and their recommendations on data collection, which have been adopted by the Airborne Law Enforcement Association and others.*
- *The Fourth Amendment's requirement that the government obtain a search warrant when the use of a UAS will intrude upon someone's reasonable expectation of privacy.*
- *The prosecution of individuals who misuse any technology to violate privacy laws - AUVSI does not condone the use of UAS for illegal surveillance. UAS manufacturers should not be held responsible for improper or illegal use of unmanned aircraft systems.*

In August 2012, the International Association of Chiefs of Police, the world's oldest and largest nonprofit membership organization of police executives, published guidelines for the safe and responsible use of UAS by law enforcement to safeguard individuals' privacy [9-5]. The guidelines include specific steps law enforcement should take to respect the privacy of individuals:

- Where there are specific and articulable grounds to believe that the [unmanned aircraft] will collect evidence of criminal wrongdoing and if the [unmanned aircraft] will intrude upon reasonable expectations of privacy, the agency will secure a search warrant prior to conducting the flight.
- Unless required as evidence of a crime, as part of an ongoing investigation, for training, or required by law, images captured by a UA should not be retained by the agency.
- Unless exempt by law, retained images should be open for public inspection.

The guidelines were endorsed by AUVSI and quickly adopted by several law enforcement organizations. Even the ACLU welcomed them to some extent. In an August 17, 2012 blog post, Jay Stanley, a senior policy analyst with the ACLU, wrote, "The IACP is to be applauded for addressing this issue, and for issuing recommendations that are quite strong in some areas" [9-6].

9.1.3 Privacy Legislation

The debate over balancing privacy and technology has been ongoing. When a new technology emerges, it is almost always met with apprehension. And as interest grew around the emergence of unmanned technology, so, too, did misperceptions of how it would be used, as well as how it could impact personal privacy. Media stories about domestic use of UAS were often accompanied by photographs of systems used in a military setting. For example, several news stories featured images or video of missile-equipped systems to illustrate stories about police systems that actually weigh only a few pounds and can only carry cameras. Headlines frequently contained the words "drones" and "spying," leaving public perception about domestic UAS skewed.

In 2012, the concerns voiced by privacy groups and civil liberties advocates about UAS technology led to the introduction of legislation restricting the use of UAS, especially by law enforcement. At the federal level, several bills were introduced in the 112th and 113th Congresses restricting the use of UAS citing privacy concerns. For example, Representative Austin Scott, a Republican member of the House of Representatives from Georgia's 8th District, teamed up with Senator Rand Paul, a Republican from Kentucky, to introduce HR 972, the Preserving Freedom from Unwarranted Surveillance Act. Representative Ted Poe, a Texas Republican, was joined by Representative Zoe Lofgren, a California Democrat, to introduce HR 637, the Preserving American Privacy Act. And Representative Ed Markey, a Massachusetts Democrat, introduced HR 1262, the Drone Aircraft Privacy and Transparency Act. Most of the legislation would require police to get a court order to operate unmanned aircraft, and some called for restrictions on even private use, such as by not allowing imagery to be taken in some public areas.

While some privacy groups hailed introduction of these bills, others criticized them and their potential negative unintended consequences. In a September 20, 2012 column in the Washington, D.C., publication *The Hill*, Tim Adelman, a legal expert on the use of UAS by law enforcement, called many of these bills "ill-conceived."

"The dozen or so bills introduced in recent months addressing unmanned aircraft and the issue of privacy not only demonstrate a misunderstanding of current law, but could have unintended consequences that hamper law enforcement officers' ability to safely and efficiently do their jobs," wrote Adelman, citing search and rescue, crime scene investigation, and surveys of multicar crashes as the types of missions the proposed legislation could needlessly prohibit. "Not only do some of these proposed bills seem to ignore established precedent, they also make the mistake of limiting law enforcement agencies' ability to use unmanned aircraft for dangerous or difficult missions that have no tie to privacy ... But under both the Scott-Paul bill, and the Poe bill, these types of UAS uses would face severe restrictions, if not outright prohibitions" [9-7].

In 2013, it was the states' turn. At least 41 states introduced legislation that in some way restricted the use of UAS, usually by law enforcement. The legislation ranged from limiting the use of UAS only in emergencies to outright moratoriums. Five states (Virginia, Tennessee, Florida, Texas, and Idaho) have actually passed legislation regulating the use of UAS, as have several municipalities and smaller government entities. Some experts worry that the legislation popping up around the country could actually do more harm than good.

"I do worry that we are missing out on the transformative potential of drones," wrote Ryan Calo, law professor at the University of Washington and noted expert on issues relating to privacy and technology, in a column for CNN. "We could see drones accompanying early morning joggers, taking sport, wildlife, and other photography to a new level, or mapping out hard-to-reach geographic terrain. The possibilities ... are endless. Which is why I would rather see an end to bad privacy law than an end to drones" [9-8]. Other countries have also pondered the issue of increased UAS use, although the bulk of their concerns have revolved around the use of armed systems and whether their countries should buy or develop them. However, in 2012 the United Kingdom's Parliament set up a working group, the All-Party Parliamentary Group on Drones, to consider the use of UAS for both military and civilian use, including the use by police and other groups inside the UK.

9.1.4 Driverless Cars

While UAS and the privacy issues they raise have received widespread attention, they are not the only unmanned system to raise questions. Automated vehicles, including self-driving cars being developed by Google, could help lead to safer roads and highways by reducing the human error behind the wheel.

In a 2012 White Paper, AUVSI found that "In 2010, more than 35,000 people died in crashes in the United States, costing the economy $230 billion per year and consuming a greater share of national health care costs than any other cause of illness or injury. Traffic congestion is an $87 billion annual drain on the U.S. economy, including 4.2 billion lost hours, or one week for every traveler" [9-9].

However, privacy advocates have sounded alarm bells, suggesting that driverless cars could track and collect data and provide it to advertisers.

"Consumers enthusiastically adopted the new technology of the Internet. What we were not told was that our use of the Information Superhighway would be monitored and tracked in order to personalize corporate marketing and make Google a fortune," John M. Simpson, Consumer Watchdog's Privacy Project director, wrote in a letter to the [California State] transportation committee. "Now that Google is taking to the freeways, we must prevent inappropriate collection and storage of data about our personal movements and environment before we allow Google's robots to take to the roads and report back to the Googleplex" [9-10].

The systems being developed include technology that can communicate from vehicle to vehicle—known as V2V—and vehicle to infrastructure, or V2I. Such vehicles can continually send out an "I am here" signal to other vehicles and obtain road information from "smart" road signs, stop signs, and even the roads themselves.

For some services, such as OnStar, being able to track a vehicle is key to the emergency response it can provide. You don't have to call for help, the system can tell that you have crashed and will send help.

"Your vehicle can also transmit important crash data to OnStar, including direction and number of impacts, rollover status, and a maximum velocity at impact. This measures the impact, and helps first responders gauge severity. An advisor shares this potentially life-saving information with responders before they even arrive," says the OnStar website.

Beyond selling information to advertisers, driverless cars' use of Global Positioning Systems—for safety and efficiency reasons—raises questions about who then has access to that information. But similar to the privacy issues surrounding UAS, this is more of a question of how information is used rather than the platform by which it was collected.

Tom Bamonte, the general counsel for the North Texas Tollway Authority, told AUVSI in a recent interview that privacy issues for automated vehicles are not a serious concern as long as they are handled in a certain way. Users don't like having their personal travel patterns shared, but they don't mind general data harvesting to improve traffic congestion and travel safety, he said. "As long as connected vehicle technology is used for general data harvesting, as opposed to tracking individual drivers, we've set a good example in the toll industry of how to find an acceptable balance," he said.

9.1.4.1 A Robust Legal Framework

The Fourth Amendment of the U.S. Constitution, which protects Americans from unreasonable searches and seizures, led to the requirement of law enforcement agencies to acquire search warrants when a search would violate someone's reasonable expectation of privacy. As the Fourth Amendment has served as the foundation for Americans' privacy rights since our nation's founding, the U.S. Supreme Court has applied its intention to evaluate new technologies in every walk of life.

In an article for the *Harvard Journal of Law and Public Policy*, entitled, "Observations from Above: Unmanned Aircraft Systems and Privacy," John Villasenor, a Nonresident Senior Fellow with the Washington, D.C.-based Brookings Institution, stated "a careful examination of Supreme Court privacy jurisprudence suggests that the Constitution will provide a much stronger measure of protection against government UAS privacy abuses than is widely appreciated. The Fourth Amendment has served us well since its ratification in 1791, and there is no reason to suspect it will be unable to do so in a world where unmanned aircraft are widely used. In addition, there are substantial statutory and common law protections that will limit the ability of non-government entities to violate privacy using manned aircraft" [9-11].

Over the past several decades, the U.S. Supreme Court has reviewed the implications of technology and aerial surveillance along with their Fourth Amendment implications and created a robust framework of precedent in which to consider how laws should be applied to UAS. The cases that provide the pillars for that framework include the following:

- ***Katz v. United States.*** In *Katz v. United States* (1967), the Supreme Court found it unconstitutional to place an eavesdropping device on a closed public phone booth, stating that the "Fourth Amendment search occurs when the government violates a subjective expectation of privacy that society recognizes as reasonable ... These considerations do not vanish when the search in question is transferred from the setting of a home, an office, or a hotel room to that of a telephone booth."
- ***California v. Ciraolo.*** In this 1986 case, the court created the "reasonable expectation of privacy" test, in which the court found that police did not violate privacy when it flew a small, manned airplane over private property and photographed marijuana growing in a backyard (within the curtilage of the defendant's home). That observation and the photographs taken were used to secure a search warrant to seize the marijuana. The Court found, "the Fourth Amendment protection of the home has never been extended to require law enforcement officers to shield their eyes when passing by a home on public thoroughfares." The Court considered "public navigable airspace" a thoroughfare and found it permissible for the police to fly anywhere the general public has the right to fly.
- ***Dow Chemical Company v. United States.*** On the same day it handed down its decision in *California v. Ciraolo,* the court found it permissible for the Environmental Protection Agency to fly a manned aircraft over a business and take detailed pictures from a sophisticated mapping camera, ruling, "The taking of aerial photographs of an industrial plant complex from navigable airspace is not a search prohibited by the Fourth Amendment."
- ***Florida v. Riley.*** The 1989 *Florida v. Riley* decision is among those most frequently mentioned in the conversation about UAS and privacy. In this case, the court found it permissible for the police to fly a manned helicopter 400 feet above a greenhouse, which was located behind a home, and use observations made by looking through missing window panels to secure a warrant to seize the marijuana being grown inside. The court affirmed its reasoning in *Ciraolo,* saying, "Any member of the public could legally have been flying over Riley's property in a helicopter at the altitude of 400 feet and could have observed Riley's greenhouse. The police officer did no more."
- ***Kyllo v. United States.*** Like *Florida v. Riley,* the 2001 *Kyllo* decision is an important case in the current debate. In the case, the court held that the use of a thermal imaging device from a public vantage point to monitor the radiation of heat from a person's home was a search within the meaning of the Fourth Amendment and thus required a search warrant. The court found that allowing the government to collect information

emanating from a house would put people "at the mercy of advancing technology—including imaging technology that could discern all human activity in the home."

Although the court discussed at length whether or not a "device is in general public use," the fact remains that when people have a Constitutional right to privacy, the police need a warrant to conduct a search, regardless of the technology being used. If it is unconstitutional for the police to use a handheld thermal imaging device from their police car on a public street, it will also be unconstitutional for the police to use a thermal imaging device to look at a house from a manned aircraft, or a UAS, absent a warrant.

- ***United States v. Jones***. In this recent 2012 case, the court found the police violated the Fourth Amendment on trespassing grounds by sticking a Global Positioning System tracker on a car to monitor a person's whereabouts for an extended period of time. The Court acknowledged that extended electronic surveillance without a physical trespass *may* violate the Fourth Amendment; however, the Court declined to rule on those grounds.

These cases and others provide the framework in which privacy and unmanned systems will be considered.

"None of the key Supreme Court precedents definitively resolve this issue," wrote Villasenor in his article. "In combination, however, these rulings indicate that the Fourth Amendment is likely to provide significantly more protection from government UAS observations than is commonly assumed" [9-11].

9.1.5 Privacy Law for All Technologies

While UAS and their potentially expanding role in our daily lives have set off a privacy debate, many observers of the issue simply see it as a new front on the general need to update privacy law across the board. As the University of Washington's Ryan Calo put it, unmanned aircraft "could be just the visceral jolt society needs to drag privacy law into the twenty-first century."

The unmanned systems industry is not opposed to addressing privacy, but it contends that the discussion should focus on how the data is handled and not how it is collected.

"If Congress wants to consider legislation addressing UAS privacy issues, it should do so in a technologically neutral way," said Michael Toscano, president and CEO of AUVSI. "Ultimately, the issue is about data. Does the government have the right to collect and use data it gathers by any sort of means against you?," Toscano asked [9-12].

Villasenor agrees that the lens through which the debate over unmanned systems and privacy is being viewed is different than how other technologies with privacy implications have been treated.

"When considering potential new statutory UAS privacy protections, it is helpful to keep in mind what has occurred with the Internet and mobile telephones, two technologies that are associated with privacy threats that are in some respects much more significant than those that will arise from unmanned aircraft. Both the Internet and mobile phones grew as fast as their underlying technologies enabled. As a result, the public and legislative dialogue regarding how best to address the privacy issues they raise has been conducted with a strong appreciation of their benefits. By contrast, while the privacy concerns associated with domestic UAS are real and deserving of attention, they are getting significant focus long before the potential benefits of the technology are widely recognized," wrote Villasenor [9-13].

The global marketplace may ultimately help settle the privacy debate. Once there are numerous examples of how this technology, whether in the air or on the ground, can save lives, concerns about privacy could lessen. As UAS are able to take to the air and help save lives, combat forest fires, and monitor traffic accidents, "good news" stories could become part of mainstream thought.

One example of such use has already occurred. In May 2013, in Saskatoon, Saskatchewan, Canada, a man who had crashed his car and became disoriented was found through the use of a small UAS equipped with a thermal imaging camera after other searches failed to locate him. By the time the man was found, he was unresponsive and suffering from hypothermia; if much more time had passed, one of the searchers told the *Star Phoenix* newspaper, "we wouldn't have been looking for a live person."

The types of markets that unmanned systems expand into could also help their public acceptance. In a study published in 2013, AUVSI concluded that unmanned aircraft will have an economic impact of more than $82 billion between 2015—the date they were scheduled to be integrated into the National Airspace System—and 2025, with the bulk of that coming from the agriculture market. This movement has already occurred in Japan, where small, unmanned helicopters sprayed nearly 2.5 million acres of crops in 2012, up from just 250,000 acres in 1995.

The use of systems for agriculture, for crop monitoring, or even crop dusting, does not raise the same privacy issues as urban reconnaissance. Likewise, statistics about reduced road accidents and deaths could help stem privacy concerns about automated cars on the road and their networking capabilities.

"We envision endless uses in the civil and commercial markets, and the process to safely and responsibly integrate them into the airspace is underway. Countries around the world are putting unmanned aircraft to use for society's benefit," says AUVSI's Toscano "Researchers in Chile are using them to study volcano plumes to understand how they move, potentially saving lives and money. In Japan, farmers use unmanned vehicles as an efficient way to spray pesticides. Right now in the U.S., legislation was just passed that charts a course for the safe integration of unmanned aircraft into the U.S. national airspace by 2015. A safe and successful integration in the U.S. could provide a springboard for the technology's development and use around the world."

9.2 References

9-1. AUVSI. Economic Report, http://www.auvsi.org/econreport.

9-2. PR Newswire, Teal Group Predicts Worldwide UAV Market Will Total $89 Billion in Its 2013 UAV Market Profile and Forecast, http://www.prnewswire.com/news-releases/teal-group-predicts-worldwide-uav-market-will-total-89-billion-in-its-2013-uav-market-profile-and-forecast-211776631.html.

9-3. AUVSI. Unmanned Aircraft System Operations Industry "Code of Conduct," http://www.auvsi.org/conduct.

9-4. AUVSI. Privacy Statement, http://www.auvsi.org/home/aboutus/.

9-5. IACP. Recommended Guidelines for the Use of Unmanned Aircraft, http://www.theiacp.org/portals/0/pdfs/IACP_UAGuidelines.pdf.

9-6. Stanley, J. (2012). "Police Chiefs Issue Recommendations on Drones; A Look at How They Measure Up." ACLU, August 17, 2012.

9-7. Adelman, T. (2012). "Flurry of 'Drone Bills' Shows Congress has Much to Learn," *The Hill*, September 20, 2012.

9-8. CNN (2013). "Bad Laws Could Hurt Good Drones," Ryan Calo, March 5, 2013.

9-9. AUVSI. (2012). "The Case for Driverless Cars," White Paper, May 1, 2012.

9-10. CNET, "Activists want stronger privacy protections for driverless cars," June 25, 2012.

9-11. Villasenor, J. (2013). "Observations from Above: Unmanned Aircraft Systems and Privacy." *Harvard Journal for Law and Public Policy*, Spring.

9-12. Senate Judiciary Committee hearing. "The Future of Drones in America: Law Enforcement and Privacy Considerations." Toscano testimony, March 20, 2013.

9-13. Villasenor , J. (2013). "Observations from Above: Unmanned Aircraft Systems and Privacy." *Harvard Journal for Law and Public Policy*, Spring.

Chapter 10
Conclusion

William Messner, Tufts University

> The automobile engine will come, and then I will consider my life's work complete.
>
> Rudolf Diesel

We hope that you, the reader, found this book to be informative, useful, and interesting. We also hope that that it convinced you that autonomous vehicles are not only on their way, but in many ways are already here with all of the opportunities, challenges, and implications for society presented by these evolutionary and sometimes revolutionary technologies.

This book covered a lot of ground (pun intended). We started with some historical context and the reasons why autonomous vehicle technologies have advanced so quickly in the last ten years: a 30 to 100 times increase in computing speed; similar increases in the capacity of solid-state memory coupled with greatly reduced costs; lower-cost, more-capable, and more-robust sensors; and the virtuous cycle of improved capabilities leading to more interest and resources applied to improving those capabilities. The chapters then covered details of different applications, value propositions, and component technologies:

- Proprioceptive and exteroceptive sensors for measuring internal states, location of vehicles in space, and obstacles relative to the vehicle have advanced greatly in the last decade, largely due to auto industry component suppliers. These advances have made autonomy in ground vehicles feasible, and future advances will eventually allow full autonomy.

- Autonomous passenger cars will have the biggest economic and societal impact of the autonomous vehicles. Replacing hundreds of millions of human-driven vehicles will be an investment of trillions of dollars with even larger benefits in terms of vastly reducing accidents, injuries, and deaths while increasing road capacities and fuel efficiencies.
- The value proposition of autonomous hauling of freight, cargo, and commercial passengers is primarily the more efficient use of limited infrastructure, which is less of a concern with passenger vehicles. The limited resources include both public roads for transit and limited-access docking facilities. However, urban buses, delivery vans, and refuse haulers, which experience many starts, stops, and turns, may benefit more from lower labor costs and reduced wear and tear on the vehicles.
- Regulations for both passenger vehicles and commercial vehicles remain a huge challenge and a huge opportunity for innovation and creativity. European countries, Asian countries, and the United States are taking quite different approaches. In America, certain states are taking the lead rather than the federal government.
- Farms are excellent places for the use of autonomous vehicles because they are benign environments compared with roads and highways. Farms have few people, lots of structure, and good terrain for use of navigation satellites, and there is huge value in decreasing the use of fertilizers, pesticides, and herbicides and simultaneously increasing crop yields.
- The capabilities of consumer robotic systems are rapidly improving from the limited (but widely available) vacuuming operations offered today. Telepresence, for example, is already becoming an application that will go beyond merely putting a video conferencing system on a mobile platform. The potential of adding manipulation and lifting capabilities has enormous implication for care of the disabled or elderly.
- Commercial and residential landscaping is an application that has much in common with consumer robots for household cleaning. There are potentially viable business cases for commercial landscaping operations that increase efficiency by allowing a worker to get off a mower to do other tasks during the mowing operation. However, at this time, the clearer value proposition is that of providing a homeowner who buys the autonomous mower the opportunity to concentrate on more enjoyable gardening and yard work tasks.
- Remotely operated vehicles (ROVs) for maritime operations are widely used for economic activities and research. The oil drilling industry is the leader in using these systems at extreme depths, and other mineral extraction industries are not far behind. Scientific and archeological research is another application of unmanned systems, both undersea and on the surface. These applications will continue to grow as oil and mineral deposits are discovered deep under water.
- Environmental monitoring is an important and growing civilian application of unmanned aircraft systems (UAS). Monitoring of tropical storms, air pollution, crops, forest health, and water resources are viable applications, as are forest fire detection and

eventually forest fire fighting. As with autonomous ground vehicles, the current and future regulatory climate for unmanned aircraft will be critical for seeing that these systems are used effectively and safely.

- Many of the same technological improvements that are making unmanned systems effective flying and driving machines are making them very good for surveillance and reconnaissance, and issues of privacy naturally arise. AUVSI and the International Association of Chiefs of Police (IACP) have issued guidelines for the use of UAS. At the same time, both federal and state legislation have been proposed, or in some cases passed. There appears to be less concern about the possibility of unmanned cars tracking their users than using UAS to monitor the public. Perhaps this is due the obvious benefits of cars to private owners, whereas unmanned aircraft are still associated with law enforcement and the military. As UAS are deployed and become more familiar, public acceptance and reasonable legislation may result.

The future is indeed bright for civilian applications of autonomous vehicle technologies and for the people who will either use the systems or who will develop them. Major car companies and suppliers of sensors are working to create these kinds of vehicles for deployment within the next 15 to 20 years. Before these fully autonomous systems are ready, it is clear that passenger cars and freight-hauling vehicles will become increasingly capable of assisting the driver. Over this same period, consumer robots are likely to become commonplace, while unmanned maritime and aerial vehicles will become increasingly capable and widely used. More of the food that we eat will have been planted, fertilized, treated for pests and weeds, and/or harvested by a piece of automated farm equipment. Last but not least, "doing the lawn" may become more like "doing the laundry," a routine task with a small amount of set–up, after which the machine does the majority of the work at the push of a button.

Index

About The Authors

Thomas Altshuler

Dr. Thomas Altshuler is the vice president and group general manager of the Teledyne Marine Systems Group, consisting of four strategic business units: Teledyne Benthos, Teledyne TapTone, and Teledyne Webb Research, in Falmouth Massachusetts, and Teledyne Gavia in Kópavogur, Iceland. He leads the day-to-day operation of the group, consisting of approximately 200 staff with full profit/loss responsibility. The Teledyne Marine Systems Group develops and produces advanced underwater communication systems, autonomous underwater vehicles, underwater gliders, and water column profiling floats, as well as inspection system for bottling and packaging industries. Prior to joining the Teledyne Marine Systems Group, he was the executive director of the information sciences division at Teledyne Scientific & Imaging, LLC, leading a world-class team of researchers developing cutting-edge technology for defense applications. In addition, he spent four years at the Defense Advanced Research Projects Agency (DARPA) as a program manager.

Dr. Altshuler received a bachelor's degree in physics from Rensselaer Polytechnic Institute and both a master's degree and PhD in physics from the Massachusetts Institute of Technology. He was a Postdoctoral Fellow at the Massachusetts Institute of Technology Underwater Vehicle Laboratory. He has received several awards, including the Secretary of Defense Award for Outstanding Public Service and the Bronze Order of the de Fleury Medal by the Army Engineer Association for inspirational leadership to the U.S. Army Corps of Engineers. He is author or coauthor of numerous publications covering a range of defense-related and autonomous underwater vehicle topics.

Vincent G. Ambrosia

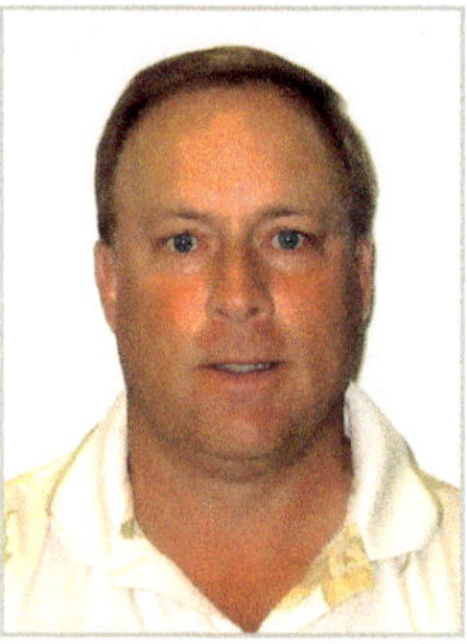

Vince Ambrosia is a senior research scientist and adjunct faculty member at California State University–Monterey Bay and is affiliated with NASA-Ames Research Center. He is the NASA Applied Science Associate Program manager for Wildfire at NASA HQ, responsible for management of a portfolio of projects related to wildfire applications and remote sensing.

He was a recipient of the 2009 NASA Outstanding Public Service Medal for efforts focused on emergency wildfire observations with unmanned vehicle systems/sensors in 2006–2009. His research team at Ames received the 2009 Federal Laboratory Consortium for Technology Transfer, Interagency Partnership Award for efforts focused on improving national wildfire observations. He has authored or coauthored over 140 papers, journal articles, and book chapters. He holds a bachelor's degree in geography from Carroll University and a master's degree from the University of Tennessee (1980). He has been employed at NASA-Ames in various positions since 1980.

Marcel Bergerman

Marcel Bergerman is a senior systems scientist at Carnegie Mellon's Robotics Institute (CMU) in Pittsburgh, Pennsylvania where he develops autonomy systems for field applications, in particular agriculture and defense. Dr. Bergerman's most recent work in agricultural robotics focused on the development and deployment of a family of autonomous orchard vehicles. The vehicles completed over 350 km of driverless operation in commercial orchards in the states of Washington, Pennsylvania, Maryland, California, and New York. Results compiled by extension educators showed that workers on the vehicles could be more than twice as efficient as workers on ladders when operating on the top part of trees. Before joining CMU, Dr. Bergerman was innovation manager at Genius Institute of Technology in Manaus, Brazil, and coordinator of the Robotics and Computer Vision Laboratory at the Information Technology Institute in Campinas, Brazil. Dr. Bergerman has published over 100 papers in the areas of robotics and innovation management and has recently coauthored a book on robust control of robotic manipulators. He received his PhD degree from CMU in December 1996.

John Billingsley

John Billingsley graduated with degrees in mathematics and in electrical engineering from Cambridge University in 1960. After four years working in the industry on autopilot design, he returned to Cambridge and gained a PhD in control theory.

He led research in Cambridge University developing early "mechatronic" systems, including a laser phototypesetting system and the "acoustic telescope."

He moved to Portsmouth Polytechnic in 1976, founding the Robotics Research Group. The results of the Walking Robot unit led to the foundation of Portech Ltd, which for many years supplied robotic systems to the nuclear industry.

In April 1992 he took up chair of engineering at the University of Southern Queensland (USQ) in Toowoomba. His primary concern is Mechatronics, and he supervises technology research in the National Centre for Engineering in Agriculture (NCEA). Three prototypes of novel wall-climbing robots were completed at USQ.

A substantial NCEA project concerned the guidance of a tractor by machine vision for very accurate following of rows of crop. Other projects have included systems for discriminating between animal species for controlling access to water, precision counting and location of macadamia nuts for varietal trials, and applications for assessing produce quality.

He is the international chairman of an annual conference series on "Mechatronics and Machine Vision in Practice," now in its 20th year. He promoted and presented Micromouse, robot maze contests around the world from 1980 to the mid-1990s.

He was awarded an Erskine Fellowship by the University of Canterbury, New Zealand. In December 2006, he received an achievement medal from the Institution of Engineering and Technology, London.

His machines have been exhibited in the Palais de la Decouverte in Paris, in the Exploratorium in San Francisco, and in the Institute of Contemporary Arts in London. Several of his robots are now on show in the Powerhouse Museum, Sydney.

Richard Bishop

A respected industry consultant, Richard Bishop focuses on the big picture and connects the right people together. He specializes in the domain of intelligent/connected/automated vehicles, including cars, heavy trucks, and buses. His clients count on him in two key strategic areas: a) to understand the complex system, technical, and societal forces in play in both the public and private sectors to enable better research investments, and b) to cultivate highly effective partnerships. He analyzes global trends analysis, develops research strategy, and facilitates partnerships and collaborations for a wide range of vehicle manufacturers, suppliers, technology firms, and government agencies around the world.

He chairs the Automated and Connected Trucking Technologies Task Force within the American Trucking Association's Technology and Maintenance Council, and serves as the Automation Subject Matter Expert for the Association of Unmanned Vehicle Systems International.

Bishop worked with H3B Media to produce a one-hour TV documentary, *The Thinking Car,* which is available for online viewing at SnagFilms.com. The documentary, funded by the European Commission, has been broadcast to audiences around the world, educating the public about crash avoidance technologies now available on many of today's cars. He also authored *Intelligent Vehicle Technology and Trends* (Artech House, 2005), which provides an extensive overview of intelligent vehicle systems.

In 2007, Bishop teamed with Ibeo Automotive Systems to serve as Team Leader for Team LUX in the DARPA Urban Challenge.

From 1991–1997, he was program manager for vehicle-highway automation at the U.S. Department of Transportation Federal Highway Administration. He established and managed the National Automated Highway System Consortium, whose work culminated with Demo '97 in San Diego, successfully showcasing automated vehicle technology to stakeholders and the news media. In 1997, he founded and continues to chair the International Task Force on Vehicle-Highway Automation, comprising public and private sector representatives worldwide to coordinate autonomous and cooperative automation R&D.

Mr. Bishop holds a bachelor's degree in electrical engineering from Auburn University, a masters in technical management from Johns Hopkins University, and a master's degree in transformative leadership and social change from the TAI-SOPHIA Institute.

Brett Davis

Brett Davis is the vice president of communications and publications at the Association for Unmanned Vehicle Systems International (AUVSI), the world's largest nonprofit devoted to the advancement of unmanned systems and robotics.

He oversees AUVSI's external media coverage and edits its publications, including the monthly *Unmanned Systems* magazine; the quarterly magazine *Mission Critical*; the weekly *eBrief*; and daily news reports on the association's website.

Davis joined AUVSI in 2007 as editor of their magazine. Before that, he served one year as community content editor for Backfence.com, a hyper-local news pioneer. Prior to that, he edited McGraw-Hill's *Aerospace Daily* for five years. He began his career as Washington correspondent for the Huntsville Times, Alabama, newspaper, where he first began to write about technology and robotics.

He graduated with a master's degree in journalism from the University of Missouri and a bachelor's degree in journalism and photography from the University of North Alabama in his hometown of Florence, Alabama. Davis lives on Capitol Hill in Washington, D.C., with his wife.

Alex D. Foessel-Bunting

Dr. Alex D. Foessel-Bunting currently works for John Deere as director of the Latin American Technology Innovation Center, located in Indaiatuba, Brazil. He is responsible for developing this new Deere technology innovation center to support business growth in Latin America. In previous roles, Dr. Foessel-Bunting has led commercialization of autonomous products in Europe and technology development and planning in the United States. Foessel-Bunting came to John Deere in 2002 after graduating with a PhD in robotics from Carnegie Mellon University in Pittsburgh, with focus on technologies for autonomous planetary exploration. Before moving to America, Foessel-Bunting participated in development of defense systems for the Chilean Navy. He obtained a master's in industrial automation and a bachelor's degree in electrical engineering and business from the Pontificia Universidad Católica, Santiago, Chile. He has written several research articles, has been awarded distinctions for academic performance and patents, and has been invited to advisory boards in innovation topics in the United States and Brazil. When he is not working in innovation and technology, Dr. Foessel-Bunting works with his family ranching operation in Chilean Patagonia.

Bradley Hamner

Bradley Hamner holds a master's degree in robotics and a bachelor's degree in mathematics from Carnegie Mellon University. He specializes in creating reliable software systems for autonomous vehicles. He also writes software components for autonomous vehicles, including perception and navigation. His work has been used in such diverse applications as agriculture, automotive, cargo transport, and security patrol. After working for the Robotics Institute at Carnegie Mellon and Autodesk Inc., he is now a senior systems engineer at Near Earth Autonomy. Originally from Baton Rouge, Louisiana, Brad currently lives in Pittsburgh, Pennsylvania with his wife and daughter.

Everett Hinkley

Everett A. Hinkley is the National Remote Sensing Program manager for the USDA.–Forest Service located in Arlington, Virginia. In this position, Hinkley is responsible for providing national remote sensing program guidance and coordination to Forest Service field units throughout the United States and serves as the Forest Service remote sensing liaison to other federal and state agencies. In June 2013, he was appointed to the position of USDA senior advisor to the National Coordination Office for Space-Based Positioning, Navigation, and Timing and is the USDA representative to the Civil Applications Committee.

Areas of his focus include land cover mapping, forest change detection, wildfire mapping, and the evaluation of unmanned aerial vehicles and new sensors for land management applications.

He holds a master's degree in mapping and geographic information systems from Ohio State University and a master's degree in geology from Miami University of Ohio.

Greg Kogut

Greg Kogut is an electrical engineer who works at the U.S. Navy's Space and Naval Warfare Systems Center, Pacific. Greg has ten years of experience working with unmanned systems and autonomous vehicles, including everything from research and development to field testing. Greg has a bachelor's degree in computer science and a master's degree in electrical engineering from the University of California, San Diego (UCSD), and worked in UCSD's Computer Vision and Robotics Research Lab while a graduate student. He lives in San Diego, California with his wife and three autonomous cats.

Justin Kraft

Justin A. Kraft currently works as supervisor of the Advanced Analysis group for the Turf and Utility division of John Deere. Located in Cary, North Carolina, he is responsible for the leadership of applying virtual verification techniques across a platform of unique vehicles that include both commercial and consumer applications. In addition to these primary responsibilities, Justin is a leader of systems engineering within the enterprise and has previously held the position of systems engineer for John Deere's first commercialized consumer autonomous lawn mower. Justin has a master's degree in systems engineering from the Massachusetts Institute of Technology. His thesis created an autonomous lawn mower optimization framework using systems engineering principles. The principles focused on incorporating multiple stakeholder values, adjacent technology trends, real options, and a platform commonality analysis. Justin has worked for John Deere since 2001 and had previously worked as a systems engineer in the power generation industry. Justin enjoys spending time with his wife and daughter and is a hobbyist in model aviation.

Justin Manley

Justin Manley has been working with marine technology since 1990. He was a principal in the development of unmanned marine vehicles at the Massachusetts Institute of Technology from 1993 to 2002. Between 2002 and 2009, he provided marine technology consulting services, primarily to the National Oceanic and Atmospheric Administration (NOAA) where he was the founding chair of the NOAA-wide AUV Working Group.

In 2009, he transitioned to the private sector, joining Liquid Robotics during its start-up phase. There he was responsible for developing new commercial and scientific programs based on the energy harvesting unmanned vehicle called the Wave Glider. In 2011 he joined Teledyne Benthos as senior director of business development. He is responsible for managing and advancing a diverse product portfolio and leads the development of a networked systems approach across Teledyne Marine Systems, which is made up of Benthos, Webb Research, and Gavia.

Manley is a senior member of IEEE, vice president of government and public affairs for the Marine Technology Society, and a member of the U.S. Integrated Ocean Observing System Advisory Committee.

William Messner

William Messner is the John R. Beaver Professor of Mechanical Engineering at Tufts University, and he serves as the chair of the department. Prior to joining Tufts in September 2012, Professor Messner spent nearly 20 years at Carnegie Mellon University in the Department of Mechanical Engineering, and he held courtesy appointments in Electrical and Computer Engineering and in the Robotics Institute. Professor Messner received his bachelor's degree in mathematics from MIT in 1985 and his master's and PhD degrees in mechanical engineering from the University of California at Berkeley in 1989 and 1992, respectively. He specializes in application of controls to data storage systems, robotic systems, and microfluidics for biological research. Professor Messner was a member of Carnegie Mellon's teams that competed in the DARPA Grand Challenges and the Urban Challenge, and he worked on the Comprehensive Automation of Specialty Crops program funded by USDA. He is Fellow of the AAAS, a Fellow of the ASME, and a Senior Member of IEEE. He is an avid swimmer and a big fan of Italian cappuccinos.

Stewart Moorehead

Stewart Moorehead is the manager of Robotics Systems at the John Deere Technology Innovation Center in Champaign, Illinois. In this role, he is responsible for leading the enterprise efforts in robotics technology development and coordinating activities with the business units.

Stewart joined John Deere in 2004 to work on the autonomous stadium mower project. Since then, he has been involved in several robotics projects, including the R-Gator, Autonomous Orchard Tractor, and Autonomous Fairway Mower. His work has focused on obstacle detection and improving the robustness of robotics technology to function in unforgiving conditions found on farms and construction sites.

Stewart has a PhD in Robotics from Carnegie Mellon University and bachelor's and master's degrees in electrical engineering from the University of Waterloo.

Dr. John F. Reid

As director of product technology and innovation, Reid is responsible for John Deere's Global Technology Innovation Network (GTIN). The GTIN is John Deere's network of advanced R&D centers in Europe, India, China, Latin America, and North America, which provides the support for the role of technology innovation in business growth. Dr. Reid and his colleagues jointly orchestrate an Accelerated Innovation Process and Enterprise Technology Process to coordinate the delivery of technology innovations to the marketplace and to build capabilities to support future business growth ambitions. In previous roles, he provided enterprise support for John Deere's technology strategy in automated and unmanned vehicle systems.

He came to Deere and Company in January 2001 after a 14-year career at the University of Illinois, where he was recognized internationally for his contributions in automation and control for off-road equipment. Dr. Reid's expertise includes machine vision perception, controls, and hardware-in-the-loop design. Previously, he served as a consultant to agricultural and military industries in their effort to deploy advanced technologies into precision agriculture and systems automation.

At the university, he was research director and chairman for six completed master's degrees and 15 completed PhD students. His awards include the Japanese Society for the Promotion of Science Research Fellowship (1998), the University of Illinois University Scholar Recipient (1995–1997), and the Karl Gardner Advising Award (1997). He has received 24 patents and has published more than 186 papers, including 61 refereed journal articles.

Sanjiv Singh

Sanjiv Singh is a research professor at the Robotics Institute, Carnegie Mellon University. His recent work has two main themes: perception in natural environments and multi-agent coordination. He leads projects in both ground and air vehicles operating in unknown or partially known environments, in applications such as mining, agriculture, emergency response, surveillance, and exploration. Professor Singh received his bachelor's degree in computer science from the University of Denver (1983), a master's in electrical engineering from Lehigh University (1985), and a PhD in robotics from Carnegie Mellon University (1995). He is the editor-in-chief of the *Journal of Field Robotics.*

Frits K. van Evert

Dr. Frits K. van Evert is an agronomist with additional expertise in robotics and in sustainability measurement. Dr. van Evert is a native of The Netherlands and received a master's degree in agronomy (1988) from Wageningen University, The Netherlands. His work on modeling crop growth and soil erosion led to a PhD in soils (1992) from Washington State University. After working at ICRISAT in Niger, and at the University of Minnesota, he joined (in 1998) Wageningen University and Research Centre in The Netherlands, where he is still employed today. In Wageningen, Dr. van Evert initially worked on simulation models of crops, livestock, and farming systems and participated in the European Union (EU) project SEAMLESS. More recently, he led the development of a robot to control broad-leaved dock, as well as the development of a smart sprayer to control broad-leaved weeds in soccer fields. Another line of investigation seeks to improve robot localization by making optimal use of information from cameras and laser scanners. In current work, Dr. van Evert aims to minimize the impact of agriculture on the environment. He uses satellite imagery, unmanned airplanes, and tractor-mounted sensors to measure crop reflectance and thus better estimate the amount of fertilizer needed by crops. He participates in the EU project CATCH-C, which analyses long-term experiments with the goal to understand trade-offs between productivity, soil quality, and greenhouse gases emissions that result from farm management options. In the EU project VALERIE, semantic web techniques are used to find documents that provide farmers with precise answers to their questions about nutrient management, weed control, and other subjects. Dr. van Evert also collaborates with the Netherlands Ministry of Economic Affairs in the development of methods to measure the sustainability of agricultural systems, with the goal of making recommendations to improve underperforming systems.

Eldert J. van Henten

Professor Eldert J. van Henten is a professor of biosystems engineering and has been head of the Farm Technology Group at Wageningen University, The Netherlands since 2005. In 1987, he obtained his master's degree (cum laude) in agricultural sciences at Wageningen University with a specialization in horticulture. In 1994, he obtained his PhD in agricultural and environmental sciences at Wageningen University, with a PhD thesis entitled "Greenhouse climate management: An optimal control approach." His research interests include sensing, modeling, design, and (optimal) control of biosystems, biorobotics, and high-tech automation and company logistics. In the field of robotics, his research activities focus on robotic capabilities like field and farm planning, robot localization and object (fruit, animal) localization, monitoring of plants and animals, robot motion planning and control, coverage planning, task and work space specific kinematic design of manipulators, and task specific design of grippers. Applications cover the domains arable farming, livestock farming, and protected cultivation and include, for example, autonomous navigation of farm vehicles, precision weed control, autonomous harvesting of fruits and vegetables, post-harvest handling of products, and the collection of floor eggs in laying hen production systems. Dr. van Henten is coordinator of the Field Robot Event, an international student robot competition in an agricultural context. He is (co-)author of one book, two book chapters, and more than 100 papers in peer-reviewed journals and conference proceedings.

Steven S. Wegener, *In Memoriam*

Steven S. Wegener passed away on August 14, 2013, in Fairbanks, Alaska, in the midst of an adventurous road trip, exploring the wild frontier of Alaska with two of his brothers. He was a senior scientist at NASA Ames Research Center from the early 1990s, serving as a science mission manager supporting high-altitude atmospheric research and later airborne science missions from unmanned aerial vehicles. Although officially retired from NASA in 2004, he never left original research, continuing his work as a scientist at the Bay Area Environmental Research Institute at NASA-Ames and with his own consultancy, Wegener Airborne Science. He served in the U.S .Air Force 1968–1971. He also worked for the U.S. Geological Survey (USGS) from 1978 until he moved to NASA in the early 1990s. At the USGS, he worked on seismicity experiments throughout the United States and abroad, in locations such as Mount Hood, Oregon, New Madrid, Missouri, Saudi Arabia, and beyond. While at NASA-Ames, he authored or coauthored over 30 papers, technical mission reports, book chapters and journal articles concerning airborne science and, more recently, unmanned aircraft systems platform uses. He is the recipient of numerous NASA Achievement Awards, and his many years of efforts will never be forgotten by his colleagues or those whom he touched with his warmth, laughter, and zest or for exploring the next scientific "adventure" or exploring innovative new concepts and places. His legacy is to live every day to its fullest, which he did right up until the end; he will forever be admired, missed, and loved.

Dan Williams

Dan Williams has worked for over 25 years in engineering and planning positions of increasing responsibility in passenger car suspension and commercial vehicle steering business units. He was originally hired by a joint venture between Moog and Lotus to develop active suspension systems. That business was purchased by TRW, where he remains, having managed the development and launch of chassis control systems in North America and Europe. Dan is currently chief engineer of advanced engineering for TRW Commercial Steering Systems in Lafayette, Indiana. He has published over 70 papers and patents on various vehicle control systems and is a Visiting Assistant Professor at Purdue University.

Dan received the bachelor's (1985) and master's (1987) degrees in engineering from the University of Illinois; his thesis topic was optimal control of racing vehicles. He received the master's (1990) business degree from the Florida Institute of Technology and then received his PhD (1995) in mechanical engineering from Florida Tech. His dissertation compared practical and theoretically optimal active automotive suspension systems. He later completed his master's (2008) in history from Purdue University, writing a thesis on the effect of the Morrill Act on engineering.

He is a registered professional engineer and a member of Tau Beta Pi, Phi Kappa Phi, the American Society of Mechanical Engineers, and FarmHouse Fraternity. Dan has organized technical sessions for SAE International and sits on various technical and standards committees, including the Truck and Bus Council. He has twice received the SAE Excellence in Oral Presentation award and the Florida Tech Outstanding Alumni Achievement Award. Dan delivered the 2012 SAE Buckendale Lecture on "Multi-Axle Vehicle Dynamics."

Brian Yamauchi

Dr. Brian Yamauchi is a principal robotics engineer at iRobot. He has 25 years of experience conducting research and development in robotics. His research interests include autonomous exploration, mapping, navigation, learning, and control for mobile robots. He currently works for iRobot's Technology Organization developing advanced autonomy capabilities for future consumer robot products.

Dr. Yamauchi was the principal investigator for the DARPA-funded Dynamo Project. The Dynamo Project developed techniques for the real-time learning of nonlinear control behaviors for mobile robots operating at high speeds and in rough terrain. He previously conducted research in high-speed unmanned ground vehicle (UGV) driver-assist behaviors for the Stingray Project, which was funded by the U.S. Army Tank-Automotive Research, Development, and Engineering Center (TARDEC). He also developed techniques for autonomous urban reconnaissance and mapping using a man-portable PackBot robot for the TARDEC-funded Wayfarer Project. At iRobot, he has conducted research and development in autonomous navigation for telepresence robots and floor-cleaning robots, as well as decentralized coordination and mapping techniques for a swarm of over 100 robots.

Prior to joining iRobot, Dr. Yamauchi developed the frontier-based exploration algorithm at the Navy Center for Applied Research in Artificial Intelligence (NCARAI) at the Naval Research Laboratory (NRL) in Washington, D.C. This technique is now commonly used by roboticists to allow robots to autonomously explore and map unknown environments. Previously, he conducted research in robot spatial learning at the Institute for the Study of Learning and Expertise (ISLE) in Palo Alto, California. He also developed a behavior-based navigation technique for a planetary rover prototype at the Jet Propulsion Laboratory (JPL) and a sensor-based control system for a robot arm to inspect the Space Shuttle orbiter payload bay doors at NASA's Kennedy Space Center. He has a bachelor's degree in applied mathematics/computer science from Carnegie Mellon University, a master's in computer science from the University of Rochester, and a PhD in computer science from Case Western Reserve University.